ENDANGERED HARVEST
CLIMATE CHANGE AND THE FOODS WE EAT

MALCOLM S. LEWIS

Endangered Harvest: Climate Change and the Foods We Eat
Malcolm S. Lewis

Cover photograph by Malcolm S. Lewis

Illustrations by Jillian Herrigel

ISBN-10: 1449577865
ISBN-13: 9781449577865

First Edition

Nature/ Environmental Conservation and Protection

CONTENTS

ENDANGERED HARVEST

INTRODUCTION
A New Pathway Sought

On November 17, 2007, at a conference in Valencia, Spain, the Intergovernmental Panel on Climate Change verified to the world once again that climate change is a serious threat to our planet. Their report brought together 2,500 scientists from 130 countries. It assembled the work that has been happening for decades to reaffirm that humans are pushing the globe into a new and warmer era. The evidence is staggering.

Atmospheric levels of carbon dioxide, the principle greenhouse gas, increased by 80% between 1970 and 2004. Secondary, but more potent, gases such as nitrous oxide and methane have also risen sharply. There is no doubt that since the time of the industrial revolution humans have been pumping tremendous amounts of these chemicals into the air. Nor is there debate that our species is to blame for their higher than normal concentrations. Their ability to trap heat is not questioned, and their effect has not gone undocumented.

Eleven of the twelve years between 1995 and 2006 were the hottest ever recorded by man. The climate is moving out of the range of natural variability and stepping into the domain of human-induced change. As we continue to pollute the atmosphere, this fact will become an ever greater reality.

Climate is one of the most basic factors that control life on earth. Changes will affect us in far-reaching and drastic ways. Our planet can look forward to melting glaciers, more heat waves, and shifts in temperature, precipitation, and general weather patterns. Faced with new circumstances, our societies will face new thresholds and challenges. There will be impacts on water availability, our most precious resource. Those living near the coasts may need to relocate. Wildfires could become more abundant. Storms could increase in number and force. Disease vectors could spread and intensify. Ecosystems, which provide an untold number of free services such as cleaning our air and water, will face certain threats. Climate change is already one of the primary drivers of biodiversity loss. As emissions continue to grow and the trends magnify, some changes might be abrupt and irreversible.

This book is an account of how climate change will affect one essential part of our lives- food. The most obvious and critical question will be, can food production meet a growing demand as temperatures grow? But that is not the only question we should ask. How can we reduce our emissions and minimize climate change through our food choices? How can different farming strategies, such as small-scale organic or industrial-scale factory farms, adapt to the changes that will come? If we want to think about both climate change and food in a serious way, all of these questions will be important.

How will climate change affect food? This book explores two specific foods that face immediate and serious threats from a changing climate. The first is maple syrup in Vermont. Without precise weather patterns, sap will not flow from a maple tree as it should during the early spring in that part of the country. Maple has a rich history in Vermont, and it continues to be extremely important to the state as a whole. Can that continue?

The second example is wine production in California. Like maple trees, wine grapes depend heavily on climate. A grape could grow in other types of weather, but it would not ripen on pace for a quality wine. And like Vermont, California relies tremendously on this agricultural product. In both places we find farmers who have built lives around their harvest, in some cases for many generations. These are not people who will easily be relocated to new states or even countries where conditions become more suitable. There is some room for adaptation. But as the climate shifts ever more drastically, adaptations will only go so far.

And what about the more basic foods that we survive on each day? On the international stage, slightly warmer temperatures could actually increase the production of many staple crops. Well, maybe. While warmer averages alone could be helpful, everything that accompanies them may bring the opposite result. As we combine those averages with higher temperature extremes and a more turbulent, unpredictable system, the crops in many regions could faces losses. And if average temperatures alone change more drastically, declines are almost certain.

Regardless of the greater trends, there will be some losers. Those in the low-latitude, dry-land tropics face the most immediate and severe threats to food security. Encompassing regions such as sub-Saharan Africa, these are unfortunately some of the poorest regions of the world. They are also regions that are home to large numbers of rural inhabitants who base their entire livelihoods on agriculture. In the United States, wheat production in the plains areas and soybean production in the southeast are both expected to fall. Even if other areas become more ideal for these crops, individual farmers could see their ways of life destroyed. And relocating this major land user puts unneeded stress on the diminishing resource of space.

To prevent these threats we must reduce emissions and slow the rate of change. This will require major commitments from nations and industries around the world. On a personal level, individuals can use food to reduce their "carbon footprint." Local foods are attached to less transportation. Organic foods come from farms that use less energy inputs and store more carbon in the soil- the world's largest carbon sink. The livestock sector contributes to 18% of the world's climate change problem. By reducing meat consumption and raising more responsibly produced, grass-fed animals, we could lower that number significantly.

Regardless of the measures we take, there will be changes, and we will have to adapt. How will food producers deal with climate change and still meet growing demand? Modern agriculture relies on uniformity and dominance of the landscape. Fewer and fewer crop varieties compose more and more of what we eat. Industrial monocultures destroy diversity and are soon vulnerable to new pests and diseases. As conditions change more rapidly, it is a never-ending race between our laboratories and natural evolution.

However there is another wave of farmers looking to more traditional strategies. They are using Organic and Biodynamic methods. Combining wisdom from both the past and present, they are demonstrating that non-industrial farming maybe can feed the world. These farmers fight pests and diseases with crop diversity and natural predators. Soil rotations and natural fertilizers rebuild the soil's health. Their philosophy is one that seeks to work with nature instead of against it. They want to prevent problems instead of treating the aftermath. In a rapidly changing system, this ideology could be the most important strategy of all.

Some evidence shows that humans first undertook farming during an earlier period of climate change, and perhaps by necessity, not choice. If the climate changes again, will our society need the enormous restructuring that it went through once before? Can we continue to grow food in a laboratory or will need to rethink our entire strategy of food production? Will we continue to ignore the threats to our environment until it is too late? Or will we find a more sustainable pathway to follow?

PART ONE
MAPLE

CHAPTER 1
Sinzibuckwud

Woksis, the great Iroquois hunter, swung his axe into a maple tree one early March night before bed. The next day he withdrew the axe and went on a three-day hunt. The slashed tree cried teardrops into a vessel on the ground below. On the day Woksis was to return, his wife found the vessel lying by the tree. Thinking it was water, she used the liquid to make a stew. When Woksis returned the stew was giving off a wonderful smell. He tasted and loved the stew, and he kissed his wife to pass on the sweet taste to her.

Nanabush, a Native American changeling, returned to his village one day to find it silent and empty. He knew that his people loved maple syrup, so he went to the woods to find them. They were all lying under the trees with their mouths open, drinking the flowing syrup. They didn't hunt, fish, or plant crops; they ate only the syrup and they were fat and lazy. Angered, Nanabush changed himself into a giant, retrieved water from the lakes, and sprinkled it over the trees so the sap would run thin. This forced the people to gather and chop wood, build a fire, and labor for the sweet luxury.

-Common Native American Legends

The Algonquin tribe called it *Sinzibuckwud-* "drawn from wood." The Cree called it *Sisibaskwat.* Native Americans slashed maple trees to collect the sap in hollowed out logs, ceramic, and birch-bark containers. It was usually the women who took hot stones from the fire and placed them into the containers to boil the sap. To keep it from spoiling they continued the process until the watery liquid turned to solid sugar.

The work was difficult, but worth it. Sap will drip from a maple tree only at the end of the winter as the weather is warming into spring. In the northeast that was a time of year when supplies ran low and food was scarce. Maple sugar nourished the Native Americans and later the European settlers in the difficult season.

Some tribes would leave their villages for weeks to go to sugaring camps. They consumed most of their sugar themselves and traded the surplus. The Ojibwa tribe, which consisted of 1,500 people, apparently made 90 tons of maple sugar one year and consumed most of it themselves as a substitute for salt.

A French Missionary named Sebastian Rasles learned to make maple sugar from Native Americans around the year 1690. He might have been the first European settler to master the practice. For farmers, there was little to do on the land during the maple season. *Sugaring,* as they called it, caught on quickly.

European settlers created their own method to boil the sap by using a series of three iron kettles. The first was the biggest and the hottest, and they transferred the sap to the smaller kettles as it thickened and condensed. They knew the sugar was ready when they could dip a looped stick into the pot and blow a long bubble with the liquid. From there, it would turn to sugar as it cooled.

The settlers referred to the first full moon of the sugar season as the *sugar moon* or *maple moon.* To celebrate the first boil they held "sugaring off" parties. These festive gatherings brought plenty of alcohol, music, and fun, as they signaled the end of winter and the start of spring.

Even with iron kettles, making maple sugar was still hard work. In 1874 a man named E. A. Fisk publicly stated

With nothing to shelter him from storms; let the wind blow and fill the boiling sap with ashes and dust, and his eyes with smoke, let him mount his snow-shoes and bring in all his sap to the boiling place upon his back, and if he finds poetry in it, I think he will say, I prefer prose hereafter.

Yet for some, maple sugaring was changing from a family hobby into a business, and their methods reflected the heightened production. Hand-carved wooden spouts replaced axe cuts. In 1860 Eli Mosher patented the first metal spout. They used an auger to drill small holes in the trees, and through them the sap dripped into metal buckets. Sugar makers learned to favor the warm side of the tree for tapping; they said that on a good day sap would run at two drops per heart beat, filling a 16 quart bucket in eight hours. They even placed small metal roofs over the buckets so they wouldn't fill with falling snow and debris. Those with larger operations abandoned their snowshoes for horse-drawn carriages to carry them through the woods. They collected the sap in larger containers to bring to the boiling shed.

In 1872 H. Allen Soule revolutionized maple sugaring with what was likely the first ever "evaporator." Soule's device rapidly increased the speed and efficiency of the boiling process. Sap enters on one side of an evaporator and gradually works its way through a maze of shallow metal pans, becoming thicker and thicker as it goes. When ready, the sugar maker releases, or "draws off," the finished product on the opposite end. And by that point sugar makers weren't just drawing off hot sap that would harden into sugar; they were also enjoying maple in the liquid syrup form.

Besides being bigger, hotter, and faster, the evaporators of today have changed little since Soule first conceived the specialized stove. One area of sugarmaking that did continue to improve was sap collection. It was only a matter of time before tractors replaced horse- drawn carriages. Even so, the idea of returning to each bucket time after time was still unappealing to many big producers. Around 1925 some sugar makers tried using a system of metal tubing to bring their sap from the trees to central collection locations. However, this was clumsy and difficult to manage and it never caught on. Then, with the advent of modern plastics, the idea of tubing rose to power again. Modern sugar makers have webs of plastic tubing spreading through their forests. They can now collect the sap from thousands of maple trees, once the work of a whole tribe of Native Americans.

*   *   *

Maple sugar was essentially the only sweetener grown in the United States until the introduction of cane sugar in the 1800's. It was Columbus who first brought the cane sugar to the new world in 1492. During his initial voyage across the Atlantic Ocean he stopped at the Canary Islands of Spain and took some on board. Yet for many years this alternative was rare and expensive in North America. Providing a new country with its sugar fix became a lucrative business for some maple sugar makers in the Northeast United States. George Cary was the owner of the Cary Maple Sugar Company in St. Johnsbury, Vermont. People referred to him as the "Maple Sugar King." He advertised and sold his sugar all over the country. His trucks drove through the streets of Los Angeles and other far-away cities to sell it to the passersby.

Over the years more and more cane sugar continued to creep into the market. Ben Franklin promoted massive maple sugar production to make the country less dependent on foreign sugar. In the early days of the colonies the maple producers could satisfy the would-be nation's sugar-fix. But as the new country expanded, they couldn't keep up. Maple sugar could only be made in the northeastern states, and people everywhere wanted something sweet.

Thomas Jefferson believed that every farmer should have at least two things- maple trees and apple trees. He had maple trees planted at his Monticello home in the hills of Virginia. Unfortunately for Jefferson his trees would not produce the sweet sap that he had found in the northeast; there was one problem- it was the wrong climate.

Cane sugar began to come from the slave plantations in the Caribbean and the Southern United Sates. By 1850 there were almost 245,000 slaves working on cane sugar plantations in Louisiana alone. As transportation got cheaper, this sugar began to creep north, even into maple territory.

Maple still held a strong advantage in the north on moral grounds. Around the time of the civil war, northern abolitionists rejected cane sugar produced by southern slaves. Of course that argument wouldn't hold up long after the confederates surrendered at Appomattox.

In 1885 lawmakers dropped the tariff on imported cane sugar and prices dropped. It became maple sugar that was rare and expensive. Yet the maple producers in the northeast were still making one thing that cane sugar people were not- syrup. Just before the turn of the 20th century maple syrup became more profitable than maple sugar. Syrup was now the "liquid gold" of the northeast. But the sugar industry that had given flavor to the founders of the United States and the Native Americans long before them had nearly vanished.

CHAPTER 2
Morse Farm

May is almost over and the sky threatens to rain. The maple season ended months before, and when I drive up to Morse Farm the grounds are quiet. The lit window in the building next to me seems, at first, to be the only sign of life.

I walk into the gift shop and a woman with short, gray hair and a wool sweater covering a turtle-neck shirt stands at the desk. She smiles and greets me and I tell her that I'm here to see Burr. She lifts the telephone receiver to call up to the house, and within minutes Burr appears to greet me with a firm handshake. He is wearing jeans, a t-shirt, and a red hat bearing nothing but the logo of a maple leaf. His full, white beard, round stomach and cheery smile give him the look of a man who has lived through many long winters on a New England farm.

Burr is the seventh generation of Morses to be a maple farmer in Vermont. His son Tom, who now works on the farm with him, makes eight. Burr's actual name is Harry I. Morse Jr., but his older sister Susie called him "Burr" when she couldn't pronounce "baby brother," and it stuck.

Burr's ancestors helped to settle central Vermont in the 18th century. The first such family member, James Morse I, came to Cabot, Vermont in the late 1780's from Massachusetts. Like many Vermont settlers, he was probably sugaring at that time, and like all of the settlers, ". . . he probably learned it from the Indians" Burr tells me.

Whether James Morse was sugaring or not, his son moved the family to Calais, Vermont, close to Montpelier, where he was definitely tapping trees. For the next 150 years James Morse's descendents boiled maple sap at that location, until Harry Morse Sr., Burr's father, moved the farm to Montpelier when Burr was five. It was from that same age that Burr had been helping his father to collect sap, milk cows, and do everything else that needed doing on the farm. Now running the business, farming is Burr's life's work.

"I've just put my heart and soul into this" he says.

About two hundred yards above the gift shop is Burr's brown-painted house. He lives there with his wife and mother-in-law. Tom lives in another house on the property- farther down the road but just within sight. Behind Burr's house is 250 acres of what most people might only see as empty woods. But among the trees and shrubs are numerous maples- the raw material for Morse Farm Maple Sugarworks. Maple farmers call this a "sugarbush," or sometimes, "sugarwoods." The only thing special or different about a sugarbush is that it's a place where farmers actually harvest maple syrup.

"People think they're gonna come and see this huge factory with trees that are just in rows" says Burr.

Those people are surprised to find the sugarbush in the state that it is. In reality the sugarbush is completely natural and completely wild. And Burr doesn't know of a single maple farm that's really much different. He tells me he has never heard of a farmer planting maple trees with the hope of harvesting the sap years later.

Walking through Burr's sugarbush I see many trees besides the maples; yellow birch, hemlock, black cherry, and beech trees are scattered throughout. Like many places in Vermont, the maples still dominate. The sugar maples are tall and straight, and their bark is a light, grayish-brown color. The ones I find aren't huge, and I can wrap my arms around many of them. The trees are not crowded, but they still manage to block the sky and turn the gray morning a shade darker. There is a light mist and the air is cool. It's the birds chirping somewhere above my head that remind me that it's still springtime.

Ferns and fallen branches lie across the forest floor. The only thing unnatural about the entire setting is the tubing that surfaces from the undergrowth in a spider-like web. In shades of green, blue, and white, the tubes zigzag from maple tree to maple tree to collect the sap. Small tubes lead to bigger tubes, and eventually to collection stations. In 1975 Burr's family laid out this network of plastic to harvest the sap from their sugarbush in a more modern and efficient way.

The destination of the trail I am following seems to be only the woods itself, and eventually the trees absorb my path back into their leaves and branches. Still, the green tubing continues off into the distance, and before long it starts to fade into the bright green leaves of the trees. If I look far enough away I can't even tell it's there. I turn and walk back towards the gift shop so Burr can show me the rest of his maple operation.

*   *   *

Tapping a maple tree to withdraw the sap inside is not always simple. Sugarmakers must consider where, how, and especially when to drill the small hole into the thin layer of sapwood.

Tap holes will scar the tree, so maple farmers need to move the holes each year and let the tree heal itself. On a piece of maple furniture that was once tapped for syrup, you might find a small round scar with a vertical tail on both ends where the tree was re-growing.

A standard tap hole is $7/16^{th}$ of an inch, but Burr has switched to a $5/16^{th}$ inch plastic "health spout." This micro-spout was developed in Vermont and it reduces the stress on the trees. In the virgin forests of the northeast Maple trees were even more plentiful than they are now. Native Americans and the European colonists would rarely tap a tree less than two feet in diameter. With more people and less trees those days disappeared. Farmers started placing more taps in more trees to gather as much sap as possible. A 10 inch tree would have one tap, a 15 inch tree would have two taps, and a 25 inch tree might have four or more taps in the same year.

"We used to put five in a big ol' tree" says Burr.

Then farmers realized that with so many tap holes the trees were unhealthy, and they began to cut back. Today Burr's biggest trees will get three taps, many get two, but most get only one tap each year. Using the health spout, Burr will tap a tree as small as 8 inches in diameter. Totaled up, Morse Farm has about 3,500 tap holes from which to draw sap.

For those 3,500 tap holes to flow, the trees need the right weather. Maples will produce sap in the late winter and early spring when the temperatures are cold at night and warm during the day. For a good "run" of sap, a maple tree needs a night of below-freezing temperatures, down to about 25° degrees Fahrenheit, and a day that warms to about 40°. When the temperatures are below freezing, carbon dioxide gas builds up inside the maple tree. As the gas warms and expands it builds pressure, which causes the sap to circulate and force its way through the small hole in the trunk.[*]

The high pressure in the tree is critical for the sap to run, and it is the difference between day and nighttime temperatures that causes it. Those temperature patterns don't exist in many places, at least not for very long, but in Vermont and elsewhere in New England during the early

---

[*] The idea behind the forcing of the sap from the tree is fairly simple, but the "sap-flow mechanism"- the pattern of the sap's movement inside the tree at any given time- is highly complex. Sap, the tree's food, moves up and down through the roots, trunk, branches, and stems, and while scientists can describe the process they do not fully understand it.

spring the conditions are ideal. Without those temperatures, the maples will sit idle and the tap holes will remain dry.

Because temperature patterns are the only thing that will produce a good run of sap, knowing when to tap the trees has long been the most important decision of all. If a sugar maker tapped his trees too early the hole would dry up and start to heal before the season started. Hammering into a tree that was too cold and hard could split the wood. If a sugar maker tapped too late, they missed part of the season and had low production for the year. Judging when to tap took a lot of experience, and some luck.

New equipment, such as airtight plastic tubing and health spouts are making this process much easier on sugar makers.

"These days with tubing you can tap early" says Burr.

In an airtight system the tap holes won't dry up and heal if left for too long before the sap starts to run. Burr told me he'd known some sugar makers tapping as early as November or December. That way they're sure to catch the whole season, even if it starts in January. But historically speaking, it isn't until March that the sap should flow.

*   *   *

Most people today use the sap from a maple tree to make syrup, and not sugar as Native Americans and early colonists once did. Even so, the term "sugaring" has stuck. Sugaring has come a long way since the days of hot stones and iron kettles, but it still involves a lot of boiling. The sap that Burr collects is more or less sweet water. In reality it is water, with about a two-percent sugar content. Burr and other sugar makers must boil most of that water out of the sap to be left with syrup, which leaves only one gallon in forty behind. The rest turns to vapor. During boiling, sugar houses fill with warm, maple scented steam. Even on a cold night, a boiling shack at a maple farm is like a sweet-smelling sauna.

From outside the Morse boiling shed appears weathered and worn. But its appearance is a stark contrast with what lies inside. Against the back wall of the shed is Burr's evaporator- a large, stainless steel apparatus for boiling tens of thousands of gallons of sap each year. When I see the evaporator it has already done its work for the season, and it sits clean, quiet and ready. It is taller than I am, and it runs almost the length of the shed- perhaps 10 feet.

Boiling out forty gallons of water for each final gallon of syrup requires a tremendous amount of heat energy, and the evaporator was designed for speed and efficiency. The large, flat pans and winding channels increase the heat exchange and speed the evaporation of the water. Burr can change sap into syrup in a mere two hours. That process would take all day using iron kettles.

Wood is the traditional fuel for boiling, and even today many sugar makers burn cord after cord of firewood each year to power their large evaporators. In certain circles, wood is considered the best fuel because of the flavor it contributes to the syrup. But many sugar makers have now made the switch to oil or gas. Burr uses a woodchip gassifier; the large round structure sits next to the evaporator and covers the remainder of the back wall. As the name suggests, the gassifier uses woodchips for fuel, but it's not the same as burning firewood.

"It's totally different than lighting a match under a pile of logs" says Burr.

The gassifier runs on woodchips that are still green, and inside of a closed system these chips produce plenty of smoke. Then the gassifier burns that smoke for an efficient, double-burning process. Connected to the gassifier, the evaporator's draft pulls the heat under the boiling pans to keep the sap steaming from start to finish.

The plastic tubing in Burr's sugarbush leads the sap directly to the holding tank in the boiling shed. When he has enough sap, he simply turns on the evaporator and makes syrup. A toilet-style floating valve automatically lets more sap into the evaporator as needed.

"Some old Yankee though of this system a hundred years ago," says Burr, "where the flue pan did the muscle boiling and then the front pan was a maze where it was forced to travel like a river."

The flue pan, also called the heat exchanger, is a big, flat, open container. Its job is to boil off about half of the water, and requires most of heat energy to do so. Then the sap moves into the front pan and winds its way back and forth as it continues to move towards the end, getting thicker and thicker on the way. The sap is constantly displacing that in front of it.

"The sweeter has to go ahead," says Burr.

Knowing when the syrup is finished boiling is more precise than the old method of dipping a looped stick into the liquid and blowing a bubble. Syrup has an exact consistency, and sugar makers must draw it off at precisely the right time. If they finish it too early the product could spoil; if they finish it too late the sugars could crystallize. If it gets too hot it could burn and taint the flavor. To know when his syrup is done Burr uses a thermometer. Like anything, maple syrup has a regular temperature at which it boils. Syrup boils at 7° F above the boiling point of water, and water usually boils at 212° F. But depending on air pressure and moisture those temperatures can change, so sugar makers often need to adjust their finishing temperature above or below the 219° baseline. Some sugar makers have automatic draw off valves on their evaporators. Burr just waits and watches. When the temperature is right, he turns the handle and hot, fresh maple syrup comes out.

Before it's ready to eat Burr filters the syrup to remove "sugar sand"- the concentrate of organic acids and minerals in the sap. The finished product has a sugar content of 67%, and its amber color shows no resemblance to the clear liquid it began as.

The United States Department of Agriculture (USDA) has a grading scale for syrup, but Vermont has a scale that's a bit stricter. Grading is based on color and consistency. Sugar makers judge their syrup from how much light can pass through it. Each year they receive a grading device from the Vermont Department of Agriculture that has five small glass jars. Four of those jars contain a syrup-like substance to represent the four standard grades: Grade A Light Amber (or Fancy), Grade A Medium Amber, Grade A Dark Amber, and Grade B. The fifth jar in the center is where the sugar maker can put a sample from their own batch. They hold it up to the light and compare it with the others.

Nothing the sugar maker does in the process can affect the syrup's grade; that is something the tree decides. In most years the trees will produce lighter grades early in the season and darker grades near the end, though that's not always the case. Light syrup comes from well-preserved sap, while dark syrup has had more contact with natural bacteria in the tree. The bacteria give dark syrup like Grade B more nutrients and a stronger flavor. I ask Burr what his favorite syrup was and what he uses it for.

"Good vanilla ice cream topped with fancy maple syrup" he tells me. "It's just really good."

*   *   *

Elsewhere in the boiling shed I can see evidence of the other side of Burr's business. Along with making syrup he uses the building to entertain groups of tourists who come, wanting to see and learn about maple sugaring. In front of the evaporator are rows of benches facing a white screen, and a digital projector hangs from the ceiling. When visitors arrive Burr shows pictures to accompany an informational talk.

Outside of the shed there is more entertainment. Various wooden sculptures that Burr has carved are set up depicting maple-related scenes. There is a carved wooden head with a white beard that looks almost like Burr; it sits on a body of a man dressed in corduroy and denim who is working at an evaporator. There is another man cutting logs with a saw. In the entrance to the boiling shed there are two sculptures that hint at Burr's political views. One is the head of an elephant. Burr chose the piece of wood because it looked like an elephant already, with wide ears and a trunk sticking down to the floor. He added eyes and a mouth, and then stuck on horseshoes for eyelashes. After he had the elephant he decided he better put something else up too. It was the trunk of a cherry tree and an apple tree wound around each other. After adding a tail, it looked just like a donkey- from the rear.

Burr's prized sculpture is off to the right of the shed.

"I wanted a good looking woman, but I knew I wasn't talented enough to carve one" he says. "If I call it 'Venus de Maple' I can live with missing arms."

*   *   *

Morse Farm was not always a tourist destination. When Burr's father moved the farm to Montpelier they had milking cows along with the syrup operation. In 1960 he sold the cows to try raising vegetables. But Harry Morse Sr. was not an experienced vegetable farmer, and he needed someone else to manage that side of the business. Burr was the youngest in his family, and his brothers and sisters were all leaving the farm.

"They had all gone on to greener pastures and I was the last holdout" says Burr.

Burr wanted to do other things as well. He played the trombone and thought of trying a career as a musician. For a time he considered becoming a long-distance truck driver. More than anything he wanted to be a writer. But his father needed him on the farm, and he sent Burr to college to study plant and soil science. In 1970 Burr graduated from the University of Vermont with his degree. After graduation he came back to the farm to grow crops, even though it wasn't his first choice.

"I ended up flailing him for a long time" says Burr.

The vegetables still never became a great success. Having studied plant and soil science, Burr tells me it wasn't meant to succeed.

"This is not a vegetable farm, this is a hay farm. It grows grass, that's what it wants to grow."

With only a maple operation to sustain the farm, Burr's father had the idea of creating a tourist destination. He started bringing buses of people to the boiling shed and giving talks. They built a gift shop, and the tourism kept expanding. As Harry Sr. was entertaining the guests, Burr reluctantly did most of farm work.

"I wasn't a great person for him to be working with." says Burr.

Despite that reluctance, Burr began to like the farm more and more. Today he enjoys it, and he's glad he stayed through all of the years.

"I love this place, and this is where I should have spent my life."

After his father passed away Burr took over the entire business. He gave the tours, and did as much of the work as he could.

"I've literally worked my butt off building the business up. I think I've worked foolishly hard."

"It's not easy money" Burr tells me. On a farm there is always a lot to do and a lot of costs. Having tourists means Burr must hire more workers to take care of them all. There are five full time employees on Morse Farm, most of them Morses themselves. During the fall season Burr takes on up to 25 workers. In October and November Burr and his workers do most of the work to prepare for the upcoming sugar season. They set and repair the web of plastic tubing, some of which has been chewed by squirrels preparing for hibernation. Sometimes Burr finds his spouts cashed in their nests. Fall is also busy with tourists who come to New England to see the striking colors of the foliage. Fall colors are especially brilliant on maple trees, and visitors have plenty to see in the sugarbush.

For Burr, syrup helps bring tourists, and having tourists also sells syrup. With his 3,500 taps, Burr could theoretically make up to 1,750 gallons of syrup per year. In 2007 he only had enough sap to make about 600 gallons. But his market is much bigger than that. He bought an extra 6,000 gallons of syrup from other sugar makers to bottle and sell under his label. Walking with Burr into the basement below the gift shop I can smell maple everywhere. The dirt floor is covered with barrels. They look like the large, steel drums you would expect to find crude oil in, but they are instead filled with sweet, amber-colored syrup.

Burr also makes other maple products besides syrup. On one side of the basement there is a metal heating pan to make maple cream with. There he boils sap until it reaches 235° F. Then he puts it in a bath of cold water to break the grain and give it its creamy texture. Through a door in the basement I see another metal pan with holes in the bottom. It's Burr's pop-corn maker that he used to make his "maple kettle corn."

Upstairs there are a number of distinct sections. Off to the side there is a room filled with dining tables and surrounded by windows. It's there that tourists come to taste the syrup drizzled on snow or ice cream.

Towards the back Burr walks me through a small kitchen where they prepare the syrup, and then into the packaging room. The room is bigger than I had expected, but it is the staging ground for selling 40% of Burr's syrup. There are rows of shelving, box rollers, and mailing labels and grading stickers sit on the counter. When I arrive the mailing room is quiet, but during the winter holidays it can send out up to 300 packages per day. Almost all of those orders stay in the country, but 90-95% of them go outside of Vermont. The biggest out-of-state market for Burr is California.

In the store Burr sells the other 60% of his syrup. I see it in every type of container from brown-plastic jugs, to clear-glass maple leaves. And the syrup selection is only the beginning. I see maple candy in the shapes of maple leaves and hearts. There is maple cream, buckets of kettle corn, maple cured bacon, maple butter, and maple mustard. There are pancake mixes, gift boxes, cheeses, hats, and shirts. My favorite product for sale is a small paper bag with a window in front to show off the large, brown crystals. It harks back to the older days of the maple harvest- actual maple sugar.

CHAPTER 3
Maples

Maple belongs to the botanical family *Acer*. The family originated in China millions of years ago. It wasn't until sometime during the cretaceous period, which ran from about 140 to 65 million years ago, that the super-continent of Pangaea completed its breakup. Before the final separation the northern borders of North America and Asia were touching, giving opportunity for maples to spread across what is now the earth's northern hemisphere. Maples existed in their greatest numbers during the Miocene period- between 25 and 5 million years ago. When the ice age came it isolated pockets of the trees, and eventually this created various species that have evolved separately. Today there are at least 124 distinct species of maples trees scattered from China to Vermont and beyond. The one that sugar makers are concerned with is the *Acer Saccharum*- the sugar maple.

The sugar maple is also sometimes known as the rock maple or the hard maple, and it is native to North America. Full grown trees will reach heights of 100 feet or more. *The Sierra Club Guide to the Ancient Forests of the Northeast*, published in 2004, listed the tallest sugar maple known to be 151 feet. The biggest tree was 23 feet in diameter, and the most ancient was 440 years old.

Their grayish-brown bark darkens with age. Its distinctive leaf shape has five blunt lobes and coarse-toothed margins. The leaves turn from light green in the spring to a darker green through the summer. In autumn the chlorophyll degrades and they change to bright shades of yellow, orange, and red.

A sugar maple flowers in the early spring, but not until its 22nd birthday. Small, yellow-green clusters containing 8-14 flowers hang and droop from the buds. 12-16 weeks after flowering they bear a small dry fruit known as the winged samara. The samaras grow in pairs, each with an attached wing about an inch long. When they fall they spin like helicopters to the ground.

The wood inside a sugar maple is hard and white. When cut, it is prized for furniture and flooring. It can produce magnificent grains with names such as "curly," "tiger," or "blind eye." It is sought after for cabinet making, and often used for bowling alleys. When well seasoned it also makes exceptional firewood.

But the best use of a sugar maple might be to leave it standing, and to wait for the sap to run each spring. Sometimes other types of maple trees are tapped, such as the black maple or the red maple, but sugar maples will give the best and sweetest flavor.

In some parts of Asia they drink the watery sap from maple trees as a beverage. It's only in the northeastern United States and bordering Canada that people have turned the sap into sugar and syrup, mostly because it's only that part of the world that has provided the proper climate to make the sap drip each year from the wood. That is why some sugar makers and scientists are starting to worry that a warming climate could devastate the maple industry in places like Vermont.

*   *   *

The Proctor Maple Research Center (PMRC) lies down a small dirt road in the town of Underhill, Vermont. It's run by the University of Vermont (UVM), in nearby Burlington. With the maple industry so important in the state, the university has been conducting applied maple research since the 1890's.

PMRC started in 1946 when former Vermont Governor Mortimer Proctor donated the land from the Harvey Farm to UVM. And with that opportunity, UVM moved their maple research to the nearby site. The center sits on about 200 acres of wooded and open land. Between 35 and 40 of those acres are an actively managed sugarbush for maple research and development.

Arriving in Underhill, I am meeting with PMRC's director, Timothy Perkins. Tim grew up in the Northeastern part of Vermont and has been sugaring his whole life. In 1991 he earned his Ph.D. in botany from UVM, and he turned his research towards sugaring as well.

One study that Tim helped conduct at PMRC was an overview of the historical trends in maple syrup production. He and other researchers looked at the timing and duration of the sugar season across the northeast states over the past 40 years. The first significant piece of data they found was that the sugar season is now starting an average of 8 days earlier, and ending 11 days earlier after just 40 years. That not only shows that the season is moving, but that it is getting shorter as well. When the typical sugar season only lasts for 30 days, three days can be significant.

"That's a 10% loss" says Tim.

Back at the Morse Farm, Burr had been telling me the same thing.

"Our seasons have been getting earlier" says Burr. "In the old days we'd wait 'till after town meeting day which was the first Tuesday in March before we tapped"

Now, Burr and other sugar makers are making every effort to be ready as soon as they can. In 2006 some collected a run of sap in January when an early thaw hit. Burr calls sugar making "95% nature and 5% man." Simply put, nature isn't pulling its weight anymore. A shift towards an earlier season would be one thing, but the climate in the northeast isn't providing the same number of days each year that can make the sap run.

"In the last 20 years it seems as if we've had an overage of bad years" says Burr. "You used to have a bad year every once in a while, and more good ones than bad, but in the last 20 years there's been quite a few bad ones. If I had to say what caused that I'd say most of the years were bad because it was a little too warm. The night temperature didn't get quite freezing enough."

If the nighttime temperatures don't drop down to the 25° F level, the sap won't expand the next day to create the pressure needed for a run.

"You need down in the mid 20's," says Burr. "You can't accept 1° below freezing, it's just not enough. Vermont sugar weather used to be 25 at night and 40's in the day, now it's maybe 30 at night and 40's during the day. It'll either run slower or not at all."

Based on the historical evidence he's collected, and the projections for an even warmer climate in the future, Tim Perkins doesn't believe that these trends in syrup production will end any time soon.

"I would suspect that what we will see is a continuing shift in the season towards earlier in the calendar year and a continued reduction in the duration of the season" says Tim.

Other scientists in the northeast agree that temperatures are rising.

In October, 2006 the Northeast Climate Impacts Assessment (NECIA) published a report titled "Climate Change in the U.S. Northeast." The report details how since 1970 the Northeast has warmed by almost 0.5° Fahrenheit per decade. Between 1970 and 2006 that adds to a warming of 1.75° F. However, that warming is not even across the seasons. The winters in the Northeast warmed 1.3° F per decade. That change in winter temperatures is a huge margin, and it might already be affecting the maple industry.

Looking to the future, the report predicts that the warming will continue. Under a scenario of lower carbon emissions, scientists predict that the Northeast will warm by between 3.5° and 6.5° F by 2100. Under a scenario of higher emissions, it could warm by between 6.5° and 12.5° F. To put that in perspective, the lower scenario would make a summer day in Montpelier feel like a summer day in Maryland by the end of the century. Under the higher scenario, that summer day would feel like the southern part of Virginia.

As Vermont plans for those years to come, it might note that states like Virginia and Maryland don't have a maple syrup industry. Tim Perkins tells me that making maple syrup in the mountain regions of those states is possible, but not feasible on a big scale.

"You don't hear about a Virginia commercial maple syrup industry because they just don't have a long enough season" he says. "They may be able to make a few gallons here and there for home use."

If the climate for making maple syrup starts to move out of Vermont it is moving northward. It used to be the United States that produced 80% of the world's maple syrup, and Canada produced the other 20%. Now those figures are completely reversed. Today it is Canada that controls 80% of maple production. Part of the reason for that shift is because plastic tubing has made it easier to collect sap in Canada where the snow is sometimes too deep to travel through. Government subsidies in Quebec have also helped to give rise to a major industry there. A third reason might be that the ideal weather conditions are moving into their territory.

If the climate does continue to warm, Vermont and other U.S. states risk not just losing their syrup, but the maple trees themselves.

"If we see this shift in climate," says Tim Perkins, we certainly expect that the vegetation, over time, is going to change drastically."

Based on temperature and precipitation, the maple forests could essentially migrate northward. Maples do exist in Virginia, but with nowhere near the density that you find them farther north in places like Vermont.

"They do have some but not a lot" says Tim.

Tim tells me the forests in those areas have mostly oak, hickory, and pine.

"They need a cold winter" says Burr, referring to his maples. "They need a cold, long rest period in the winter, and how many years are they gonna go without it and still be happy?"

That change wouldn't happen in a few years to be sure. Trees are resilient organisms, and a few warm years isn't enough to kill them off. If the climate in Vermont starts to resemble that of Virginia, it would still take decades, maybe even a century for the sugar maples to start thinning out and disappearing. But in a new climate, change can't be avoided forever. And those changes, however far off, will be dictated by our actions and choices now.

"Eventually the forests that we have will resemble the forests that grow in those conditions" says Tim, referring to the woods farther south.

*   *   *

Even as the climate trend reaches upward, not all of the years are warmer. In 2007 the sugar season came late. For much of the spring it was too cold and Burr had low production.

"A lotta people use that as a way to come to people like me and say 'you're full of shit, you're saying there's global warming and here it is too cold on the 15th of April'" says Burr, mimicking the argument that others would make against him. "Well, I don't take it back at all, my words, because that's one thing global warming does, it gives you unpredictability with a capital 'U.'"

Year to year variability is often an argument that some will hold on to. Each time a colder year comes around doubters may try to convince Burr that climate change is a false reality and that he has nothing to fear. In truth, one season that is especially hot or cold says very little about larger trends. It says more about the less-predictable weather patterns than about the gradual and long term climate which scientists across the world agree to be on the rise.

Burr may be worried about the future, but he is happy to say that Vermont is still making plenty of maple syrup.

"We're still makin' the most in the states and the best," he says, "but to do that we're having to change our ways."

Advances like plastic tubing help sugar makers deal with the weather, and draw sap from the trees even if the conditions aren't perfect. The first advantage to plastic tubing is that sugar makers can tap their trees in advance to take advantage of the entire season, even if it comes early. More importantly, the tubing has an airtight, vacuum seal. Sap runs out of a maple when the pressure outside the tree is lower than the pressure inside. A cold night followed by a warm day will create that condition naturally, but sugar makers can help create them by taking pressure out of their tubing.

"The vacuum will counter the weather," said Burr, "you're fooling the trees into running."

But plastic tubing can only help so much. In a good year Burr should be able to make ½ gallon of syrup from each of the 3,500 taps he has in his sugarbush, for a total of 1,750 gallons. His best year so far with those taps was only 1,000 gallons, and in 2007 he had a mere 600 gallons.

It could be worse. Those without the vacuum-sealed tubing, who still use the traditional method of bucket collection, are struggling in the poor conditions.

"The bucket people are having terrible seasons" says Burr.

And Burr predicts the conditions will get worse.
"I think it's gonna be more difficult."

Sitting with Tim Perkins in his quiet office in Underhill, I ask him if he thinks that at some point there will be no sugar season in Vermont.

"Well, it depends what you mean" he says. "You don't have to get to the point where there are zero days of good flow periods for people to stop making maple syrup."

Like any business, the sugar makers in Vermont will only produce and sell syrup if they can make a profit. And like any business which depends on the weather, they can only follow a changing climate so far. Adaptation has limits, especially if the root problem is not addressed. If the conditions become like they are in Virginia now, they might make a few gallons, but it won't be a state industry.

Tim's research shows that sugar makers have lost 10% of their season over the past 40 years. It's difficult to say how much more they would have to lose before making maple syrup ends up costing more than it's worth. Plastic tubing, stainless-steal evaporators, and everything else that goes into producing syrup is a big investment, especially when it's used for fewer and fewer days each year.

"Over the next 100 years I would expect the industry in the U.S. to become less and less economically viable" says Tim. "And if the tree resource disappears that'll finish it off. Not good."

CHAPTER 4
Vermont

With only 8,000 residents Montpelier is the smallest capitol city in the United States. On Main Street the brick buildings never rise above three or four stories. Just a mile from downtown there are woods and rolling hillsides. In the large white statehouse, topped with a gold-plated dome, the government presides over the 600,000 people that live in the state.

Legislators in Vermont have been taking climate change seriously and looking for ways to reduce emissions. On September 16, 2003, Republican Governor Jim Douglas issued the "Climate Change Action Plan" for Vermont, which stated specific goals targets. It aims to reduce emissions, from a 1990 baseline, 25% by 2012, 50% by 2028, and "if practicable using reasonable efforts" 75% by 2050. It's an ambitious goal.

In 2004 Vermont emitted seven million metric tons of Carbon Dioxide, the lowest of any state in the nation.

That's partly because Vermont's population is smaller than any state's except Wyoming's. Seven million metric tons might not seem like a lot compared to the almost six *billion* metric tons that the United States pumped out in 2004, but it still beats 111 out of the 218 countries or territories that the United Nations Statistics Division listed for CO_2 emission in that same year. That includes countries like Paraguay at 4.18 million tons (population 6.1 million), Tanzania at 4.35 million tons (population 40.5 million), and Afghanistan at only 0.69 million metric tons of CO_2 (population 27.1 million).

It's true that on the grand scale, Vermont is not a major player in driving the Earth's climate into a warmer period. Many of the states or countries that do beat Vermont in carbon emissions do so in a big way. Even if Vermont's emissions went to zero, those reductions would pale in comparison to what goes on elsewhere.

"It's a worldwide problem" says Tim Perkins, "it isn't something Vermont is going to solve on their own."

Burr hopes that more action would happen on the national level.

"There's nothing little old Vermont can really do" he says.

"But it does take every individual to do their part" says Tim.

And it's the individuals that stand to lose the most in a warmer climate that should be leading that movement. Vermont has a lot to lose. The maple industry is a major part of who they are today.

The state has the highest concentration of sugar maple trees of any state in the nation. Their official state tree is the Maple. Their state quarter, printed in 2001, reads "Freedom and Unity" beneath an engraving of a maple farmer in the woods. Vermont even has an official state flavor- maple.

For hundreds of years maple sugaring has supported an industry and an element of character that has become completely intertwined with Vermont. The state's maples support roughly 2,000 sugar makers, each tapping between 50 and 50,000 trees. The Vermont Maple Sugar Makers' Association represents many of those producers statewide. Founded in 1893, it is the oldest known agricultural organization in the country.

Economically speaking, maple syrup by itself isn't an enormous financial enterprise in Vermont. The total marketable value of the product in 2005 was $11.4 million. However, when you consider the supporting industries, which produce materials such as Burr's evaporator or the tubing he uses to collect sap, researchers calculate the total economic impact on the state to be $226 million.

In every calculable way Vermont produces more maple syrup than any other state in the US. In 2005 the USDA listed Vermont as having 2,140,000 tap holes, which was 31% of all taps in the nation. In the same year those taps produced 410,000 gallons of syrup- 32% of the nation. In descending order of total production, the other states that the USDA measures as maple syrup producers are New York, Maine, Pennsylvania, Wisconsin, Michigan, New Hampshire, Ohio, Massachusetts, and Connecticut.

If you ask Burr, he will tell you that the real issue isn't how much syrup they produce or how much it's worth. The significance of the maple industry in Vermont goes far beyond that. "It's huge," he says.

In 2007 the town of St. Albans, Vermont, held the 41st annual Vermont Maple Festival. They had a road race, crafts, face painting, music, cooking demonstrations, and a 'Maple King and Queen' pageant. Farther south, in Pittsford, Vermont, there is the New England Maple Museum. There are rooms filled with maple exhibits and artifacts, followed by a slide show and syrup samples.

But it's not just Vermonters' own fascination with maple that's important. Maple is an important part of Vermont's image as seen by outsiders. Burr is on the board of the Vermont Attractions Association, which works to promote tourism to the state.

"It's no easy matter to keep people coming to Vermont" he says. "We don't have Disney World or an ocean. We don't. We've got the maple image, that's our biggest, biggest thing."

Maple is of course the center of the tourism business that Burr operates, and he believes that that maple is a big factor in all types of tourism across the state.

"There's not a single one of those people who is not thinking of Vermont maple syrup."

The NECIA report notes that many parts of the economy in the northeastern United States are driven by climate. If the ski industry- a major part of seasonal tourism in the area- starts to dwindle, it could mean fewer consumers of maple syrup and other local products.

Between seasons such as fall foliage and winter skiing, tourism adds $1.5 billion each year to Vermont's economy. If the lure of maple syrup plays a part in that tourism, as Burr claims it does, then the maple industry's financial implications are far larger than $226 million. More syrup might mean more skiing trips or more visits to the bed and breakfasts to see the autumn leaves. And more visitors also means selling more syrup.

"Part of the reason why they come, or why they like it when the do come or why they stay a day longer is because of the maple image" says Burr. "It's just part of the deal."

That's why keeping Vermont's maple image strong is so important to Burr Morse and many others in the state. With other states in the country vying for the top spot, and Canada already flooding the market with their syrup, the fight for the maple image is not easy.

"Our maple image, we gotta keep working for it. Maine is trying to takeover as the top producer; New York has been for years. Canada is huge and they'd love to takeover our image."

A few years of low production won't destroy an image that Vermont has built for literally hundreds of years. But if they want to hold onto the image for a long time into the future, they will have to keep producing a large quantity of syrup. And producing maple syrup in Vermont is something that climate change could continue to challenge.

Burr knows what he and his state are facing. He doesn't want Vermont to lose its maple image, but he knows that if the climate shifts that may happen.

"It's an image that we've had for these 200-plus years, but it's not infallible" he says.

*   *   *

A few years back the Vermont Public Interest Research Group (VPIRG) contacted Burr. They wanted somebody to testify in front of the state legislature on how climate change was affecting the maple industry.

"I've always been more of a conservative voter," he says.

And according to Burr, VPIRG is politically as far left as you can go.

"I really searched my soul, I said 'do I want to do this?' and my soul said yes."

He did do it, and since then he's continued to be vocal on the issue.

"We've gotta keep the message out there that this has affected Vermont maple sugar makers."

Even if Burr is in favor of national action on climate change, he does think that individuals and industries could help create the movement, and that the maple sugar industry could be one of those.

"We can be leaders in the movement if there can be anything done to combat it."

As climate change affects maple syrup, it also affects Vermont. It's not just producers like Burr that have built lives around the product for generations, it's the rest of the state as well. If the sugar makers suffer, so will Vermont. With a state's image at risk, the maple industry could be in a position to make a call for change; they could lead the state towards reducing emissions. And since Vermont maple syrup has recognition around the country, it could be part of a bigger, national movement on climate change as well.

Whether or not that movement takes place we will all have to decide. For now, Burr Morse and other sugar makers across the state are still making syrup and working to keep Vermont's maple image alive.

"I don't know whether there'd be something that would replace it," says Burr, "but we've been working now for 200 years on this maple image deal, so we gonna take 200 more years to come up with the next image? Let's keep the one we have."

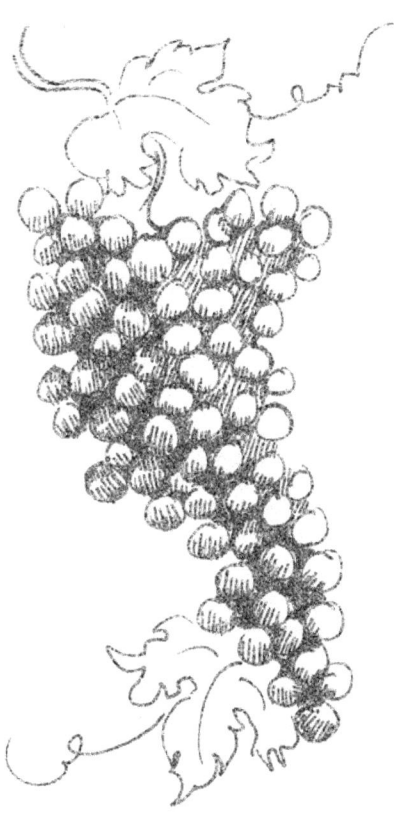

PART TWO
WINE

CHAPTER 5
Biodynamics

The first thing I notice while walking around the Benziger winery is the smell. Whether I am out among the vines, next to the compost pile, or near the fermenting tanks- it's as though I've spilled a glass of red wine across the front of my shirt. The fumes follow me everywhere.

Mike Benziger and his wife Mary bought the ranch in 1980. They searched for two years, and finally found the unkempt and overgrown property in Glen Ellen. In the first year they transformed the 85 acres into what would become a burgeoning Sonoma winery.

Growing up in New York, Mike worked for his father who imported wine and spirits. After college he moved to California and worked at the San Francisco wine shop Beltramos. In 1978 he took an apprenticeship at Stony Ridge Winery in Livermore, California, and after he found the Glen Ellen property he decided to start his own label.

Mike's parents and siblings joined him at the ranch, and in 1981 his family started creating their first wines. With the idea of selling good, affordable wine, they introduced the Glen Ellen brand in 1982. It prospered for over a decade, but in 1993 Mike sold the popular label. He wanted to produce unique, higher quality wines; he also wanted to respect the land. To accomplish both he explored Biodynamics. In 1997 the Benzigers hired the world renowned horticulturist and Biodynamic expert Alan York to help them become a certified Biodynamic vineyard.

An original thinker named Rudolf Steiner developed the concept of Biodynamic farming in the early 20th century. Steiner, an Austrian, lived from 1861 to 1925. He earned a Ph.D. in philosophy, and became an author, architect, artist, theatre director, and more. His unique philosophy he called "anthroposophy"- in some ways a mix of spirituality and science. It was this thinking that created Waldorf Education, which now includes over 900 schools around the world. Waldorf schooling takes an interdisciplinary approach to educate the whole individual. It is very much creative as well as analytical, and seeks to nourish healthy growth in children.

Like with his schooling methods, Steiner's farming ideas adopted a holistic approach. Biodynamics builds on sustainable and organic principles and incorporates the forces of nature, seeking to create a closed, self-sustaining ecosystem on the farm.

That closed system would partly explain the strong smell of wine that surrounds me. It is early September, and the grapes are being picked and crushed. During the whole process nothing leaves the farm. Stems and skins are added to the enormous compost piles. Grapes that have started to prune are left sitting on the ground to rot into the soil.

"Everything that's on the land is being used to put back on the land" says John Garn, one of the environmental consultants for the Benzigers.

John has a goatee, brown hair, glasses, and a brisk smile that permeates his cheeks and eyes. His relaxed, short-sleeved, button-down shirt seems appropriate for the day, which at 8:30 in the morning is already hotter than I am used to at my home in New Hampshire.

A major platform of Biodynamic farming is to concentrate on the plant and its environment both above and below the soil. The theory is that healthy soil will encourage deep, strong roots. In turn, that will strengthen a grape vine so the grapes can draw stronger, more intense, and better balanced flavors from the ground. In order to nurture healthy soils and plants, Biodynamic farming uses only natural, organic applications.

John explains to me how a strong chemical such as a pesticide, herbicide, or fungicide can often do more harm than good for a grapevine. It may kill the pest or fungus, but it also kills the good organisms that the plant needs to be healthy and to protect it from other threats.

"The whole thing is vulnerable now, that's the cycle we get into."

John views the problem as a grouping of microscopic chairs that surround each plant. Those chairs can be filled with helpful organisms, or harmful ones; if you use chemicals to kill things, who knows what might sit down next? Instead, Biodynamics fosters the growth of the good organisms with the hope that they fill up the chairs and leave no room for problems to occur. The farming strategy meets its goals with natural composts, fertilizers, and preparations- all of which come from the farm.

Instead of using chemicals to kill weeds, Biodynamic farms rely on cultivation. And instead of spraying for pests, they encourage helpful insects and animals to live there, forming natural predator-prey relationships.

"It's very analogous to whether you believe in western medicine, which is you get a disease and then you treat it, or eastern medicine and homeopathy, where you take things to prevent you from getting sick" says John. "Think of Biodynamics as the homeopathy approach."

In order to successfully run a biodynamic farm, the Benzigers base their farming practices on the basic principle of observation.

"Go out there and look" says John.

By spending time on the land, making careful observations, and developing a connection to the vineyard, Mike Benziger and his team work to anticipate problems. The goal is to fix problems before they occur, by eliminating the conditions that cause them.

*   *   *

The Benziger ranch is set in a complete bowl with steep hillsides. Next to Sonoma Mountain, it lies between

Sonoma Valley and the Valley of the Moon. The unique terrain was formed by Sonoma Mountain's volcanic explosions about 2 million years ago. Rich soils there include ash, welded tuff, obsidian, basalt, chalk, clay, argilla, and hardened lava and volcanic rock.

With diverse soils and terrain, no two plots are equal. Different areas of the vineyard have different slops, elevations, mineral contents, sun exposures, and drainages. All of these factors affect the vines, and therefore the flavors in the grapes. After years of studying his vineyard, Mike has divided it into 29 distinct blocks. He and his crews micro-farm these 29 blocks, treating each one differently based on the unique conditions it exhibits.

About 65% of the grapes that the Benzigers grow are Cabernet Sauvignon- frequently referred to just as Cabernet. To complement it, they have other classic Bordeaux varieties such as Merlot, Cabernet Franc, Malbec, and Petit Verdot. They also have a variety of white grapes such as Sauvignon Blanc and Chardonnay.

Cabernet Sauvignon grapes are small and spherical. The black, thick skins are resistant to damage and disease. It's that toughness, along with the popular flavor, that's led the grape to be the most successful red variety in California. In 2002 there were almost 76,000 reported acres of Cabernet under cultivation in the state, beating second-place Merlot at just over 52,000 acres. (Including white grapes, Cabernet placed second only to Chardonnay, which had nearly 99,000 reported acres.)

Cabernet emerged on the banks of the Gironde River in Southwest France. There, in the Bordeaux region, it became a popular part of local blends. Because of the French precedent, the grape is still often blended today in the United States and elsewhere. A small size and thick skin produce a high amount of tannins in Cabernet. Mixing it with varieties such as Merlot gives the wine a smoother texture.

Less than half of the 85 acre Benziger vineyard is covered by grapes. Scattered throughout are gardens, wetlands, cover crops, and wildlife and insect sanctuaries. Trees surround the property on all sides. Grapes may be the Benzigers' main crop, but they attend equally to all areas of the Biodynamic Vineyard.

Attracting beneficial wildlife helps maintain predator-prey relationships. Insect gardens bring butterflies, praying mantis, and helpful mites to protect their vines. While Bluebirds may eat some of the seeds on the property, they also feed their young on the larvae of vine pests. Cover crops can attract harmful animals such as gophers, so the farm has habitats for owls and hawks. The birds patrol their grounds free of cost; one works the night shift and one the other the day.

A highlight at the ranch is the constructed wetlands which sit at the bottom of the bowl, surrounded by vineyard slopes. Besides catching rainwater, the wetlands are here to purify all of the wastewater from the winemaking process. The system consists of two separate ponds. The first has five small machines floating on the surface, with ropes fastening them to the banks on either side. As I approach the first pond only one of them is turned on. It rocks back and forth as it churns and mixes the water below. John tells me that the machine is an agitator. Waste that enters into the pond contains biological material; when that material hits a body of water it creates 'biological oxygen demand.'

"To break it down requires oxygen."

Depending on the load entering the pond, they can turn on all or some of the agitators to mix in more or less air.

After the wastewater sits for about 5 days in the upper pond, it percolates through a bed of gravel and soil into the lower pond about 20 feet away. The lower pond is covered in a thick bed of what looks like bright, green algae, but is actually a fern. In fact, the surface plant blocks sunlight to prevent algae from growing, thus keeping the water clean.

By the time the water reaches the lower pond, about a ten day process, the purification is complete.

"The quality of this water is pretty darn close to pure" says John.

According to regulations, when the impurities in the water are at least below 97 parts per million (PPM), a vineyard can use the water to irrigate their grapes.

"The sample I saw was 7 PPM" says John- almost perfect.

Recycling that water to irrigate their vineyard saves the Benzigers about four million gallons each year. And leaving those extra four million gallons in the ground preserves it for other people and wildlife that might need it in the often dry climate.

*   *   *

After seeing the ponds I keep walking through the vineyard. Along the path there are olive and fig trees and a flowing stream. The Benzigers try to practice more traditional estate farming, the way people did a century ago. Even with a primary crop, they don't want a monoculture.

I pass a team of workers who are using hoes to weed under the vines. Each man works his way down an individual row. While the weeds are pulled, most of the grasses and cover crops growing between rows are left in place. Old methods of viticulture- or grape growing- say that other plants will compete with the vines and degrade the quality of the grapes. With the belief that other plants can boost soil fertility and control erosion, the Benzigers grow cover crops such as mustard, rye grass, crimson clover, and Austrian winter peas.

To help moderate the growth of these grasses and crops the Benzigers have a flock of sheep. In the fall and winter they graze among the vines. But in the spring and summer they are kept away from the hanging grape clusters that can prove to be a more tempting meal.

The other domestic animals the Benzigers employ are three Scottish Highland Cattle. The beasts have a thick, shaggy coat of hair, large heads, and horns a couple of feet long that curve upwards into a point. Two of the Benzigers' cattle are brown, and the third- the newest of the three- is light blond. Their pen and small barn is against the treeline on the far corner of the property; there they can lounge in the shade and eat. In addition to helping the sheep manage the cover crops, the cattle provide valuable manure to fertilize the fields.

Another important fertilizer is compost. When I walk down to see these compost piles, the mounds of weeds and pressed grapes skins loom over me. Some are perhaps double my height. The winery used to pay between $2,500 and $4,000 per month just to get that waste hauled away; now they use it to their advantage. They turn the piles four-five times each year to keep the bacteria working. When it's ready, a tablespoon of the compost can contain a billion organisms. Workers apply the compost once per year throughout the rows of vines, timing their efforts after the fall harvest but before the cover crops are drilled, to coincide with the vines' root growth.

Another way the Benzigers increase soil vitality is with Biodynamic herbal preparations. Those include chamomile, to stabilize nitrogen and stimulate growth; oak bark, to strengthen the vine's immune system; dandelion, to aid in photosynthesis; yarrow, to help the vines absorb minerals; and equisetum, to strengthen the vine's resistance to mold and fungus. As part of the closed Biodynamic system, all of the herbs for these preparations are grown on-site.

*   *   *

During my visit to the Benziger ranch it is the middle of harvest. Biodynamic farmers try to align the cycles of their grapes as closely as possible with the cycles of nature. Bud-break, when the growth cycle of the vine begins and green shoots start to emerge, should happen on the spring equinox; bloom, when the vines flower and self-pollinate, should happen on the summer solstice; harvest, when the grapes are picked, should happen on the fall equinox; and dormancy, the winter rest period, should happen with the winter solstice. Under this philosophy, the closer the plant aligns with these natural cycles, the better the grapes will express that years attributes. It is still early September, over two weeks before the fall equinox, but the grapes are ripe. I pick a Cabernet grape off of a cluster and taste it. It is juicy and sweet. I try a few more.

In the hot, dry weather, I am impressed at how green and vibrant the vines are. Similar to elsewhere in Sonoma County, the climate at the Benziger ranch has helped to grow high quality grapes. At 2,250 feet above sea level, neighboring Mount Sonoma delays the fog from the Pacific Ocean, 20 miles to the west, giving the winery more sun. In July and August, average highs range into the 80°'s and 90°'s Fahrenheit. The nighttime lows are between 50° and 55° F. Warm days ripen the grapes, and the cool nights moderate the pace of that ripening; both of which are important for well developed flavors.

The four main flavor components of wine are acidity, tannins, sweetness, and alcohol content. They can give wine taste characteristics such as being crisp, soft, or dry. In a good wine, none of the flavor components will stand out by itself. While flavor profiles are partly due to the grape varieties selected and the winemaking process, ripening is critical. Contrasting high and low temperatures promote balance in the grape's flavors.

An unripe grape is tart, but as time passes it becomes sweeter and less acidic. The skin gets thinner and the tannins become bolder and less astringent. The various subtle flavors and aromas in the grapes become richer and more complex. To achieve good balance, these changes need to line up properly. A grape that ripens too fast, for instance, in a climate that is too hot, will be too sweet, lose its acidity, and have immature, undeveloped flavors.

Mike and his crews visit each vine many times each year; their goal is to achieve uniform ripening across all of the 29 vineyard blocks. Around harvest, Mike works especially close with the grapes. Deciding when to pick is based on flavor, and that means going out and tasting. Picking the grapes too early or too late, even by a few days, can give winemakers flavors they don't want. When the grapes are picked and crushed into juice, the winemaking can begin.

At its core, winemaking is a very simple process. In fact once the juice is placed in a container, time alone can handle the rest. Natural yeasts that live on the grapes eat the fruit's sugar, creating a by-product of carbon dioxide and alcohol. Unless the container is closed, the carbon dioxide evaporates, and fermented wine is left. Of course, to consistently make *good* wine, the process is more involved.

Winemakers control many parts of the fermenting stage including the size and material of the fermenting containers, at what temperature the grapes ferment at, and how long the juice stays in the containers.

At the fermenting station I see enormous stainless-steel vats. They are perhaps 20 feet high and 10 feet in diameter. Pipes go in and out to fill and drain the containers of grape juice. Other containers are smaller, and I can peer into them to see the crushed grapes soaking in their skins.

While some winemakers add yeasts to help the wine ferment, Biodynamic farmers like Mike Benziger let the process take place with only natural yeasts that are living on the grapes. After fermenting for 10-12 days, the Benzigers leave the wine in the containers for almost another month. Rising carbon dioxide brings the skins to the surface to be separated from the juice. Then they leave some wines in the containers to age and mature; others they drain into oak barrels where the wine can sit for almost two years. If made of French oak, the 60 gallon barrels can cost up to $700 each.

At Benziger they store their barrels in an underground cave, which they built by boring a system of tunnels into the hillside. The cave totals 28,000 square feet and holds 4,000 barrels. It maintains a naturally cool temperature of 63° F with high humidity. A few times each year, on the full moon, they go through the cave and "rack" the wines- draining the barrels and then cleaning out the sediment before refilling them. According to Biodynamic thought, the full moon is important because its stronger gravitational pull may help to compress the sediment and further clarify the wine.

When the flavors have matured the wines are ready to get bottled. Some varieties, such as the Bordeaux style wines, are blended. Others remain separate by grape varietal, or even by the specific vineyard block it came from. They may continue to age in their bottles for another year before Mike will release them for sale.

*   *   *

The Benzigers began their transformation towards Biodynamics in 1997, and it wasn't until 2001 that they obtained the strict certification from the Demeter Association. Following the principles laid down by Rudolf Steiner, Demeter exists in 20 countries and certifies Biodynamic farms around the world. But Biodynamics was only part of Mike Benziger's goal, and in 1997 he had also hired John Garn to help him focus on the bigger idea of sustainability.

John graduated from Sonoma State University in 1981 with a degree in Environmental Planning. After a brief career building massage tables in Sweden, he moved back to northern California. He was working with high-risk youth, and then he started an environmental division for his friend's organizational development company. Eventually John opened his own company as a green business consultant. In the early 1990's he began by helping small business owners be in compliance with environmental regulations. The community gave out a recognizable sticker so the public knew who was certified. He started with automobile repair shops, and then moved

on to wineries. Besides Benziger, John now consults for
many vineyards in the surrounding towns.

The first place that John looks for sustainability is
on the farm, which can mean far more than simply having
an organic or even Biodynamic certification.

"There are a lot of ways that you could *be* organic
and still not be sustainable" says John. "Your practices
could suck, you could have erosion problems, you could
overuse the chemicals that you're allowed to use. There's a
perception with the public that organic is the highest level.
And in reality, you can be a farmer who is certified organic
and still be a bad farmer."

Even though Biodynamic regulations are strict, the
Benzigers can't claim to be fully sustainable. With inputs
that include electricity, tractors, and even human labor,
their vineyard is not always the closed-loop, self-sustaining
system they might strive for. So can a place like Benziger
ever be sustainable?

"I am a firm believer that there's not a destination
called sustainability" says John. "It's a direction, not a
destination. You're always headed in a more sustainable
direction."

In some cases, moving towards sustainability can
mean tough decisions. By eliminating synthetic herbicides,
for instance, Mike and his crew might spend more time on
a tractor to eliminate weeds.

"It's tradeoffs. Each location is radically different."

For those reasons, some people in the industry have
come to feel that some synthetic chemicals might even be
better for the environment if it means you can reduce your

fossil fuel consumption. Those types of questions can often make the best path towards sustainability difficult to find.

John is helping Mike to be more environmentally sustainable throughout the winery, not just in the vineyard. They have already analyzed energy and water conservation and solid waste recycling, and they are moving on to environmental purchasing. They want to look at everything that comes onto the property. To start with, they are researching the top five product inputs for each department. In administration, for instance, this includes copier paper. Is the paper they buy recycled? Has it been shipped from far-away locations? Did it arrive in recyclable packaging?

Gathering such information is the first phase in moving forward.

"It's giving them a feedback" says John.

John's sustainability audit covers more than just environmental decisions, too. In order to be a sustainable business, he believes you need to consider things such as relationships with neighbors and the community. And of course, a critical part of sustaining any business involves economic profits.

Just like out in the vineyard, they can face tough decisions, some of which might come when they compare environmental sustainability to the sustainability of their checkbook. But at least they will have the information they need to make an educated decision. The better they understand what's behind the actions they are making, the better they can decide what steps to take.

"They can look at it and make sense of it," says John.

*   *   *

I ask John Garn about Benziger's image and the market for their wine.

"I think they've been seen sort of as the radicals, the wild guys, doing things that are non-conventional" he says. "But they make fantastic wine."

And as a private, medium-sized, family operation, the Benzigers can experiment with their wines in any way they choose. As opposed to a large winery that needs uniformity in its product, Benziger offers more unique flavors.

"They can do all kinds of quirky things. They have more flexibility."

John tells me that between 50 and 60 employees work at Benziger. With that staff they're able to make 120,000 cases of wine each year. A savvy wine drinker might notice that an 85 acre estate won't produce anywhere near 120,000 cases of wine, especially when only of 40 of those acres are used to grow grapes. Like Burr Morse in Vermont, the Benzigers have a good market for their wine label, and they buy grapes from a close network of about 50 other growers. Throughout the morning I see three or four truckloads come to unload huge bins of fresh-picked grapes.

Since Benziger buys so many of the grapes from other growers, the vast majority of their wines are not certified as Biodynamic. Only the estate bottled wines, made with grapes farmed on the property, bear the small Demeter logo. For the rest of the growers, Mike and his consultants have created a sustainability program of their own called Farming for Flavors.

"He doesn't want to be buying from people that aren't part of the philosophy that he wants to follow" says John.

The Farming for Flavors program focuses on the broader idea of sustainability that John promotes in his work. When Mike proposed the program to his growers, only a few of them decided to opt out. To officially certify the majority that remained, Mike enlisted the help of the Demeter Association.

A central part of the program is a farm plan for every vineyard. This includes a map, a pesticide use report, and an action plan for what each grower plans to work on in the coming year. Every spring the Demeter Association compares the action plan to actual results, and evaluates the vineyard accordingly.

Farming for Flavors also mandates that the participating wineries take part in California's Sustainable Winegrowing Program. The program, which began in 2002, is a product of the California Sustainable Winegrowing Alliance, a non-profit organization based in San Francisco. Supported by the Wine Institute, and the California Association of Winegrape Growers, they publish a voluntary, self-assessment workbook for growers in the state. Its vision is to make wine environmentally sound, socially equitable, and economically feasible. With 14 chapters and 227 criteria, the audit covers issues such as soil management, ecosystem management, energy efficiency, solid waste reduction, water conservation, air quality, environmentally preferred purchasing, and neighbor and community relations. After completing the audit, wineries get back a customized report showing how they compared to other wineries of similar size.

The last part of Farming for Flavors is the feedback meetings. The growers come together to discuss what is and isn't working at their vineyards and to taste different wines.

Mike and John are still in the process of developing a point system for the program. By becoming more sustainable, growers will receive more points. And following the theory that ecologically and socially responsible grapes can make better wines and attract more business, Mike will pay growers more based on the point level they rise to. But just as when the Benzigers transitioned to Biodynamics, switching to more sustainable practices can take time and effort.

"There's a risk factor involved" says John.

By paying the growers more, Mike is helping to share that risk.

I ask John if it is harder for the grape growers to incorporate the various sustainability practices into their vineyards.

"It depends on the attitude of the grower. A lot of guys like to use chemicals because they're quick and easy" he says. But, he continued, "We've only been chemically dependent for about 50 years."

John believes firmly that to stop using those chemicals requires "not only the technological component, but more importantly the organizational and behavioral changes that go along with that."

Like at the Benziger ranch, growers need to spend more time among the vines to be able to recognize early signs of pressure. And because the owners can't tend each vine themselves, to be successful they need a team of well trained workers. Those are steps that require a real commitment. The Farming for Flavors program can't transform each vineyard overnight, but it will hopefully get them pointed in a more sustainable direction.

CHAPTER 6
Wine's Challenge

It's difficult to say exactly when winemaking began. For certain it was thousands of years before today. The only technology people needed for the discovery was a container to hold grape juice. By just letting it sit around they would soon notice the natural changes the juice went through, even if they didn't understand the cause. And after they noticed how the beverage affected their heads, they probably made a habit of letting the liquid ferment.

The first real evidence of serious wine production comes from about 8,000 years ago. The area is part of modern-day Iran, in the region surrounded by the Caucasus Mountains, the highlands of eastern Turkey, and the Zagros Mountains. At the time, the climate there was likely cooler and more humid- hence more suitable to grape growing.

From Iran, the tradition of winemaking pushed its way through the Middle East, Egypt, Greece, Italy, and into Europe. During the colonial period, that same tradition spread all over the world.

However, colonizers didn't have to start from scratch in the new lands. Winemaking grapes fall under the genus *Vitis*, and by the time seafaring Europeans reached the other continents, *Vitis* was already there. Many foreign peoples had been making wine on their own for hundreds if not thousands of years. But, what the Europeans did bring was their own specific species of grape, *Vitis vinifera*.

Native grape species in the Americas survived in common use for some time. *Vitis rotundifolia* and *Vitis labrusca* were two examples in Mexico and the United States. It was Franciscan missionaries, using native grape species, who brought winemaking north from Mexico to California. Moving along the coastline, they founded missions to make sacramental wine. By 1769 the missions had reached San Diego, and by 1823 they totaled 21 in number. In 1831 the Bordeaux native Jean Louis Vignes, one of the missionaries, started planting *Vitis vinifera* instead of native varieties.

Agoston Haraszthy, a Hungarian Count, brought many new European varieties to California. He began in the southern part of the state before moving to Sonoma in 1857. John Downey, California's governor at the time, commissioned Agoston to transplant transplanting the grapes to help expand what was by then a growing industry. New vineyards continued to open, especially in the northern counties of Sonoma and Napa. Agoston founded Buena Vista Vineyards in Sonoma.* He also wrote about wine and improved methods.†

It wasn't long before the rest of the world's grapes became second class to *vitis vinifera*. Today *vinifera* dominates winemaking; it includes Cabernet Sauvignon, Pinot Noir, Chardonnay, Riesling, Zinfandel, and all other grape names we see on most labels. Wines made from other grapes do exist, but they are about as rare as the wines made from other fruits such as blackberries or peaches.

Even many of the *Vitis vinifera* grapes are highly uncommon. Within the species there are perhaps 10,000 varieties. At small wineries, especially throughout Europe, you can find winemakers using some of these unique grapes. Many others go untouched for decades. Only about 50 makeup what most people drink.

On the heels of the California gold rush, the state's wineries experienced huge growth in the 1860's and 70's. Then, in the later years of the 19th century, disaster struck. Wineries in both the United States and Europe were attacked by phylloxera, a small louse. Phylloxera fed on the roots of vines and destroyed them. Native to the Americas, the bug found no resistance on the new European species of grapes. *Vitis vinifera* was threatened with extinction anywhere that phylloxera reached.

Unable to eradicate the louse, winemakers discovered that they could graft their European grape

---

* The Winery is now called Buena Vista Carneros
† In 2007 the Culinary Institute of America inducted Agoston Haraszthy into the Vintners Hall of Fame.

varieties to North American roots without affecting the grapes' flavor. Today there is still no cure for phylloxera, and resistant rootstocks from the Americas are used nearly everywhere.

Saved from phylloxera, winemaking in California soon faced another threat. On January 16, 1920 the eighteenth amendment to the United States Constitution went into effect- prohibition. For the nearly 15 years that followed, any legal production of wine disappeared. But, on December 5, 1933, amendment 21 gave the industry a new hope. Its words "The eighteenth article of amendment to the Constitution of the United States is hereby repealed" were a savior to winemakers and to alcohol supporters of all kinds across the country.

The wine industry took some time to regain momentum. During the first thirty years after prohibition California mainly produced cheap wines. To make matters worse, winemakers copied the European system of regional labeling as if those regions were literally types of wine. A California Bordeaux might use the same grape varieties and taste similar to a Bordeaux from the actual French region, but it wasn't from France.

In 1966, winemaker Robert Mondavi left the Charles Krug Winery, his family's business, and started a new winery under his own name. The Robert Mondavi Winery in Napa began making premium California wines, and named them after grape varietals rather than European regions. The rest of California followed Mondavi's lead, and production took off.

California is by far the country's leading wine producer, accounting for about 90% of the industry today. And the United States ranks fourth in total wine production around the world, behind third place Spain, and second and first place competitors Italy and France.

Mondavi's success is also part of the reason why Napa County rose to such prominence. It is home to almost 400 bonded wineries, and it is still the most famous wine region in the nation.

Sonoma County, home to the Benziger ranch, is directly to the west of Napa. Sonoma is over twice Napa's size, but comes in second with 260 wineries. It is also second in fame, though it has some big names such as Gallo, Sebastiani, Korbel, and Kendall-Jackson.

Across California there are now 2,275 bonded wineries. With their 477,000 bearing acres, they produced 543 million gallons of wine in 2005. Compared with the 63 million gallons the remainder of the United States produced, California dominates. In retail value, wine is California's primary agricultural product. The 441 million gallons sold in the United States alone brought in $16.5 billion. But the impact on the economy doesn't stop there. Tourism to wine country is up to almost 20 million visitors a year, adding another $2 billion to the state's cash flow. When considering all of wine's supporting industries, such as restaurants, bottle making factories, warehousing, and more, California wine adds 309,000 jobs in the state, and another 875,000 jobs in the rest of the country. In total, the impact of California's wine is $51.8 billion dollars in California, and $125.3 billion in all of the United States.

For an industry that faced huge setbacks at the end of the 19th century, and again at the start of the 20th, this is no small feat. But California's wine industry could be facing a new challenge- a warming climate.

*   *   *

Each year in January the American Society for Enology and Viticulture* and the California Association of Winegrape Growers join forces to hold the Unified Wine and Grape Symposium in Sacramento, California. With a combination trade show and seminars from the industry's leading experts, it is the largest function of its kind in the United States. Seminar topics range from vineyard practices to micro-oxygenation to marketing. At the 2007 symposium a new topic emerged- climate change.

Dr. Greg Jones, a climatologist from Southern Oregon University, spoke during the first seminar on Climate Change, giving an overview of the issue and its possible effects on wine.

"I don't have to tell anybody in here the very tight connection between climate and wine, that's pretty clear" said Jones.

For wine grapes, excessive heat can mean a low quality grape that's too sweet, and not enough heat can mean an acidic grape that never reaches maturity. In California's wine region, the climate gives a good balance of hot and cold and the grapes mature well. But with such

---

* Enology is the study of wine. Viticulture refers more specifically to the study and practice of grape growing.

a high reliance on the favorable conditions, the wine industry faces a greater threat.

"Compared to broad acre crops- corn, soybeans, things of that sort- that narrowness of the climate structure and suitability really puts this industry at much greater risk" said Jones.

Thus far, the climate in California has usually done its part. However, scientists and winemakers are already starting to note that there is a shift underway. Compared to the earlier years of California wine, there are noticeable differences.

"It's clear we've seen changes in average climate structure and variability, absolutely without a doubt."

In his work Jones looked at the temperature changes in Napa Valley between 1930 and 2004. During that time period, the average minimum temperatures rose 5° Fahrenheit. And correspondingly, the number of days below freezing dropped in all seasons. The day of the last spring frost moved 45 days earlier, and the day of first autumn frost moved 42 days later, creating a growing season that is now three months longer. And those changes were not gradual, much of it happened during the last 24 years leading to 2004.

The warmer winters have helped many growers in some ways.

"We've seen warmer dormant periods" said Jones, "and for some people this is actually a good thing."

Jones noted that a warmer winter could reduce the frost damage on vines. But from another perspective, a cold winter can help kill off pests and organisms that live on the crops. Some winemakers even rely on the cold temperatures to freeze their grapes while still on the vines, using them to make sweet dessert wines.

"How about the ice wine harvest that didn't happen this year in both North America and Europe- real big issue."

In March of 2006 the California Climate Change Center published a white paper titled "Climate Scenarios for California." To make their predictions, the groups chose what they believed to be the best suited computer models available. They predicted that the average temperatures in northern California would rise between 2.7° and 8.1° Fahrenheit by the year 2100. They also predicted that the temperatures could rise by up to 11.5° Fahrenheit in the summer- which is when the grapes grow and ripen.

Looking to the future, Jones told his audience that the temperatures during the grape growing season and the ripening periods would continue to lengthen. Maximum temperatures and the number of hotter days would both go up. He said that the average scenarios suggest a 15-25% increase in the number of days suitable for grape growing. In effect, this will also open up more area for wine. But he knows that counties like Napa already have a long, warm growing season.

"Is there a threshold, do we need to be concerned with that?"

To make good wine, the important factor isn't just how many days are available to grow grapes, but whether or not those days will grow quality grapes. A report in the

Proceedings of the National Academy of Sciences in 2006 examined how Climate Change might affect wine production in states like California. The report, of which Jones was an author, was titled "Extreme heat reduces and shifts United States premium wine production in the 21ˢᵗ century." As is clear from the title, the results were not good.

The report agreed that with average temperature increases, the total area for growing grapes could rise. But what the authors found to be more important than the average temperatures were the extremes. When they included extremes in their models, the areas that were only marginally suitable for wine production were wiped out by 2100. The area that could support the growth of premium quality grapes declined by over 50%.

"Major, major reductions" said Jones at the symposium. The only areas that still could grow high quality grapes moved towards the coast, up in elevation, and north.

These reductions are due to the way the grapes ripen. In hot weather, the sugar ripeness might come early while the ripeness for flavor and maturity would still come later. With climate change, it could essentially be both too early and too late to harvest the grapes on the same day. So while areas like Napa and Sonoma could see more land open up for their vineyards, they may just end up growing a lot of bad wine.

After Greg Jones there were two other presenters on Climate Change. The first was Dr. Robert Wilkinson from the School of Environmental Science and

Management at the University of California, Santa Barbara. Wilkinson, who's research focuses on water policy and climate change, talked about those issues at the symposium.

Different computer models show almost every conceivable possibility for how rainfall will change in the region. With climate change, California's grape-growing regions could get a lot wetter or a lot drier, and nobody knows for sure.

"That doesn't give a lot to work with" said Wilkinson.

But since rain and moisture obviously play a critical role in the growth of any crop, he suggested that farmers start to consider how they would deal with whatever situation might come their way.

If the conditions get drier, farmers could join the rest of the state with big shortages. With growing demand and a shrinking supply, some people will start to get cut off.

"In California every major water supply is already over allocated."

One major water source is in the mountains of California. At the moment, the winter snow and ice accumulation gradually melts, giving many places in California the water they need year-round.

"We're getting free water storage," said Wilkinson.

If the temperatures go up, and that snow turns to rain, people lose that valuable service. It could cause more winter flooding and a loss of total water availability.

And more rain for growers could be just as bad as less.

"What happens if we get rain in August or September?" asked Wilkinson.

For a grape grower, a lot of rain at the wrong time is just as bad as having none at all. Many growers like Mike Benziger try to practice deficit irrigation- challenging their vines with slightly less water than they want. This keeps the grapes smaller and gives them more intense flavors. Too much water when the grapes are maturing would lead to watery, bland wine.

The good news, according to Wilkinson, is that most of the water use in California comes from smaller sources. It's taking care of those smaller sources that will matter.

"Between ground water and local projects, that's most of the water supply" he said. "Managing groundwater is going to be very important."

He suggested that by being more efficient, especially in urban areas, California could respond much more efficiently to shortages.

Following Dr. Wilkinson's presentation was Dr. John Trumble, an entomologist[*] at the University of California, Riverside. In Dr. Trumble's talk he outlined a study he conducted to see how plants, and the insects that feed on them, would react in an environment with more carbon dioxide.

---

[*] Entomology is the study of insects.

Trumble and his researches initially believed that they would see an increase in plant productivity with elevated CO_2 levels. And they did. They doubled the CO_2 levels, and the plants responded well- they got bigger by about 25%.

"We were getting about a 20% increase in overall photosynthetic activity" he said. "This actually increases the amount of leaf area on the plant."

The plants grew faster, but the insects also ate more. While plants are based on carbon, insects are instead based on nitrogen. More carbon dioxide in the atmosphere was making the leaves of the plants bigger, but those leaves weren't drawing more nitrogen from the soil. Therefore, there was less nitrogen in every square inch of leaf. To get the nitrogen that they needed, the insects had to eat more.

Increased temperatures that go along with climate change will also generate more insect growth. Because most insects can't control their own body temperatures, they respond dramatically to external conditions.

"They feed and develop and reproduce directly in proportion to temperature" said Trumble. "If you increase the temperature by 3 or 4 decrees Celsius [5.4-7.2° F], you're going to have population explosions in insects like we've never seen before."

Plants, and the farmers who tend them, will surely have to find ways to deal with more insects eating more leaves.

"If you can't control those insects you suffer a major economic loss and reduction in those plants."

Another issue that Dr. Trumble discussed was how the plants responded internally to elevated carbon dioxide levels. Because the ratio of carbon to nitrogen shifts, it upsets the balance that plants have adapted to over thousands of years. With elevated CO_2, plants become stressed for nitrogen.

"Many growers are going to likely begin to put on more nitrogen" said Trumble, referring to fertilizers.

And while it can temporarily help a single crop, more nitrogen isn't always the best answer for the environment or the farmer. Not only does the production of nitrogen fertilizer require an energy-intensive, industrial process, but nitrogen runoff pollutes waterways. Nitrogen could also negatively affect the grapes. Dr. Trumble was researching the leaves, not the grapes themselves, but he does know that adding more nitrogen will increase phenolics in the plant- one of the major components affecting the wine's flavor.

"There are a large number of uncertainties to what we might know about the future" said Dr. Greg Jones at the end of his talk. "But I don't think that uncertainties should put us in a quagmire of not doing anything."

And Jones wasn't only talking about California or the Nation, but wineries too.

"This industry doesn't have an immense part in the footprint, but it has enough of one that we need to talk about it."

And winemakers won't be talking about their carbon emissions just because they feel an obligation to the rest of the world. In a changing climate, with a sensitive crop, they themselves have a lot to lose.

"Even if the climate models have it 10% right, 50% right, we still have an issue" said Jones.

*   *   *

At the Benziger ranch, they are noticing a changing climate in some of the same ways the scientists had talked about.

"We're finding ourselves in the early phase of transition to something that we're not familiar with" says John Garn.

How exactly the changes are affecting their Biodynamic vineyards, their grapes, and ultimately the wine they sell, they don't yet know. More than anything, they are trying to learn to adapt to what comes their way.

From what John tells me, the biggest change they fear is in extremes, just as Dr. Jones had said.

"There seem to be bigger fluctuations in heat and rain" says John. "If we go from having three or four triple digit days in a row to two weeks of triple digit days in a row, it changes the whole way you manage your vineyards."

He gives the simple example of canopy management. Growers typically trim the leaves to improve air circulation and reduce mildew. With an unexpected heat spike those grapes can literally sunburn and dry up.

"And you can't put the leaves back on or little umbrellas out there."

The Benzigers are also seeing bigger fluctuations in harvest times than they have before. They are accustomed to a swing of perhaps five days or a week; their harvest times are now arriving up to two weeks off schedule. It's hard to know how that unpredictability is affecting the grapes.

Even the basic rules that grape growers once followed may not apply in a warmer climate, and new lessons will have to be learned.

"You have to ask yourself the question 'well what does this really mean?' and I don't think people really know yet" says John. "Your whole planning concept has to change."

If the climate does keep changing, some grapes might fare better than others. In hot weather, it's mostly the white grapes that are more susceptible to damage. That's good for growers like Mike Benziger who have a majority of red grapes, but even reds could face challenges.

"There's a limit to how much superheat they can take" says John.

Some grapes might adapt to warmer temperatures over time. However, finding the vines that adapt the best can be difficult. Since winemaking is a business, farmers can't just let their grapes grow wild and see which of them survive in new climate conditions. Luckily, that's where science can step in.

"I think what you're gonna find is researchers playing with different clones and rootstocks."

Even if growers try to switch to more heat resistant grapes, they will still face hurdles. They will have to plan ahead for warmer temperatures and predict what will work years in advance.

"You can't say 'oh next year I'll rip those out and plant a better varietal.' You just can't. Because it's a five or six year cycle before you can get any fruit," says John.

So grape growers will have to know what changes are coming, and when. If they start to fall behind during the climate's transition, they could find themselves with a year or two of low production, poor quality, or both. And like any business, they need to produce a good crop each year to make money.

"You only have so many seasons to try things," says John.

CHAPTER 7
Better Business

A changing climate will bring challenges to winemakers like Mike Benziger, but it also gives Mike another perspective from which he can analyze his business. With climate change in the picture, there is a new impetus to examine energy use, whether it is from electricity or tractors or anything else on the farm. Like any other input in the Biodynamic system, fossil-fuel based energy is something Mike wants to minimize.

John and the Benzigers are doing a complete audit on carbon emissions. As with the other sustainability measures they are implementing, they first want to get the facts. They are totaling up the figures for oil, natural gas, and even the airline travel that takes them to conferences or marketing events.

"How much carbon are we responsible for?" says John. "Once they get the inventory number, then they'll be able to set a target."

Along with that target they are also creating a plan to reach it. The Benzigers are trying to approach the issue in a strategic way, finding the best and most efficient ways to reduce their carbon footprint.

"In the world of adaptation strategy it's called 'no regrets.'"

They believe that there are many ways to decrease their emissions and increase their profits at the same time. John tells me that they are realizing they might be better off making efficiency gains than investing huge sums of money in a solar array, for instance. Being more efficient is proving to have a faster payback- saving them more money, sooner. And with less energy demand, when they do make an investment in something like solar panels they will need less of them.

The Benzigers want to utilize the information they gather for more than just their own benefit. As they learn how much energy they're using, they are also calculating how much energy that means per case, bottle, or glass of wine. Along with other measures like water usage, they are starting to put this data on fact-sheets and make it available in their wine shop or on their website. Customers and other winemakers will be able to view that information. If other wineries produce the same facts, they can start to learn from each other.

"Until we get those metrics, we can't have a conversation within the wine sector" says John.

As producers observe each other using less energy or water per case of wine, and therefore saving money on those inputs, the industry as a whole can start to improve.

*   *   *

Even if winemakers and other industries around the world do improve efficiency and lower emissions, some degree of change is already coming. If growers want to survive, they will have to use the right strategies and be quick to adapt.

In the closed system of Biodynamic farming, how well can growers like Mike Benziger adapt? Because they can't use quick solutions from synthetic chemicals, some might argue that their risks are higher if the temperatures rise. But Biodynamics could also work to Mike's advantage as conditions change.

Like elsewhere in California, one issue the Benzigers might face is water.

"The wells here are going down" says John. "Even after heavy rain years the wells still go down."

If businesses and homes face tighter water allocations around Sonoma and Glen Ellen, the Benzigers will be at an advantage. With their recycling ponds, they already save four million gallons of water per year.

"You're in a better position now if climate change comes and reduces the water amount."

On the other hand, if the rains increase, those ponds at the bottom of the vineyard will help conserve the runoff. And with more cover crops and grasses growing among the vines, they will have less erosion problems during heavy rain events.

If insect populations see rapid growth, the Benzigers won't be able to add chemicals to quickly dispel the problem. But they can try to stay ahead of the problem with their insect garden and predator birds. Neither can they add chemical fertilizers to fix carbon and nitrogen imbalances. But they can make adjustments and increase their herbal preparations and natural manures.

The most important advantage that the Benzigers might have is their farming philosophy. Their techniques are based around careful observation and planning for the future. In Biodynamics, they constantly must work to prevent future problems before they occur, not just as they happen. And adapting to problems like climate change will need to involve thinking ahead.

Like any winery, the Benzigers will face risks. They can put in hours of observation among the rows of grapes, and they still won't predict every change that will come. But all farmers will face setbacks. In the end, it could be the mentality of adaptation that gives some growers the edge. And with a community of 50 growers, all working together to find sustainable solutions, the force behind that mentality becomes even stronger, and the possibility for successful adaptation seems even more real.

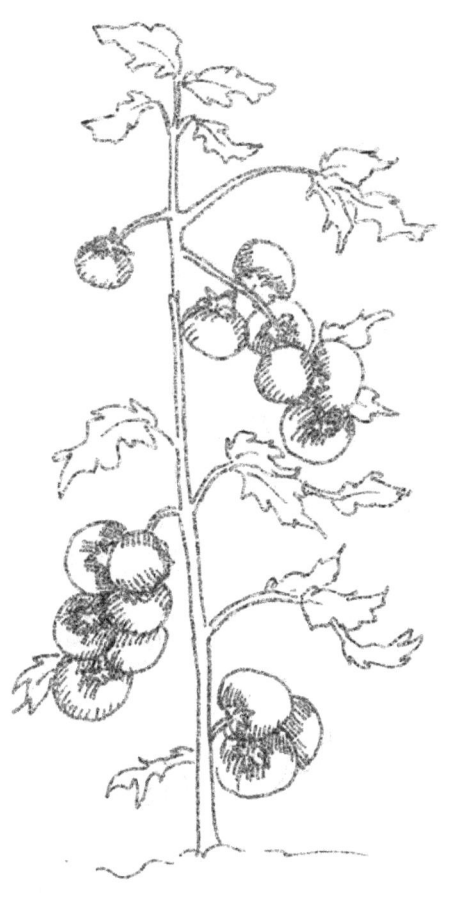

PART THREE
FARMING

CHAPTER 8
Dayspring

The hardest part about digging potatoes is finding them. Steve works down the row with a shovel to loosen the dirt, and Julia and I follow on our knees, pulling the small red tubers up by hand. If I reach down far enough and get a good grasp on the plant, many of the potatoes emerge still attached to the stems. It is the stragglers I can't seem to locate. I push my hands down into the dirt as far as I can and feel around where the plant just was, but I never seem to collect as many as the more seasoned farmhands.

Steve, Julia, and Zach, are interns at Dayspring. They are there for the summer, and they help the farm run during its busiest season. Charlie Maloney, the farm's owner, recruits interns from the nearby College of William and Mary, in Williamsburg, Virginia. It's there that he teaches a class on sustainable agriculture in the environmental studies program.

We pick the potatoes from their stems and throw them into buckets. The rest of the plant, and any particularly ugly potatoes, will stay behind; along with the spuds that evade me, Charlie will plow everything back into the soil.

Under the hot Virginia sun the field is dry, but down near the potatoes some moisture remains. After an hour of digging the dirt is caked over my body. The loose, white clothing I chose, with the idea that it might keep me cool, is soon the same color as everything else- brown. In three days it will be the solstice, the start of summer, and the temperatures are well into the 90's. We rest often, stepping into the shade of the barn and drinking water.

On the 18 acre property in eastern Virginia Charlie and his family grow mostly vegetables, along with some fruit and cut flowers. I have never been a farmer, and as I walk around the fields during this early part of the season I can't tell the difference between a pepper and a squash plant, unless I can actually see the food hanging from the stalk. Around his rectangle-shaped farm Charlie points out cabbage, eggplant, lettuce, chard, okra, blackberries, strawberries, beats, cucumbers, onions, squash, garlic, sweet potatoes, green beans, melon, and much more. Some plants, such as tomatoes, have many different varieties. Even Charlie isn't sure exactly how many plant varieties he has growing in his fields, but perhaps around 300.

The farm is dotted with structures of every shape and size. The large, white house sits near the road. Looking behind it there are small sheds for storage and for washing and sorting vegetables. There are high-tunnel greenhouses for some crops- which are especially important during the cooler months. There is the chicken coop, the horse and goat barn, and the pig pen. The largest barn gives Charlie storage and room to work indoors. Off on one side of the barn his son Paul has assembled a workshop for his motorcycle repairs. To the other side there is a small kitchen and a bedroom where the interns live. In the back there is a room where the Maloneys hold a Quaker meeting each Sunday. I am not sleeping in the barn for the week, but instead in a small camping trailer, located on the tree line at the far end of the fields. Behind that the only other structure is the tree-house the family built years before.

Charlie's grey hair is tied into a pony-tail, and he is balding on the top of his head. His full beard is grey as well. He wears a t-shirt, shorts, and tall leather boots. With many days spent in his fields, his arms are tanned brown. He has strong legs, and walks with the cantor of a man half his age.

Born in 1950, Charlie has been farming his whole life. His parents farmed and raised their family on the eastern shore of Maryland. There, he and his four older siblings helped to manage the 200 acre property where they were tenants.

Charlie describes his parents' farm as "kind of old-fashioned, diversified farming"- something he says was typical of the era. They tended to a small dairy herd, chickens, turkeys, tomatoes, cucumber, corn, soybeans, and more.

During his senior year of high school Charlie met Merriam. She was a year below him. They would marry in

1973, after Merriam graduated from Western Maryland College.[6]

"We started dating and just kinda kept going" he says.

After high school Charlie went on to college as well. He studied at Duke University, and graduated in 1972 with a major in religion. Following that, he attended Duke's Divinity School for a master's degree, which he earned in 1976. That same year he was ordained as a minister for the United Church of Christ. In 1978 he acquired a second master's degree at Duke in Theology, with a focus in pastoral counseling and psychotherapy.

After finishing his studies, Charlie moved to Virginia with Merriam. He became licensed as a professional counselor in the state, and he began to practice pastoral counseling and psychotherapy. They moved to Hampton, and then to Williamsburg as they were beginning to start a family.

Charlie enjoyed counseling, but he still maintained a desire to work the soil. Regardless of where they were living, he and Merriam were always raising some of their own food.

"We always had a garden" says Charlie, even if at times it just included tomatoes- his favorite crop.

But they were looking for more. They wanted a homestead of sorts, where they could farm and raise kids. In 1987 they bought their land in Cologne. They started farming more and they home-schooled their children.

---

[6] The school was renamed McDaniel College in 2002

Charlie was already selling some extra produce to friends. With his new piece of property, he expanded that venture.

Still directing a pastoral counseling center in Williamsburg, Charlie couldn't farm full time. But over the coming years that would change. The farm was where Charlie most wanted to be. While the farm expanded, his counseling practice declined.

"I gradually cut back on my work" says Charlie. "I gradually began to be here more and in the office less. As I got older I felt more drawn to give myself time on the farm."

In 2001 he left his practice for good.

"I feel a greater sense of contentment and satisfaction in this work," says Charlie of farming.

When people learn that Charlie spent 23 years working as a pastoral counselor, they wonder how he made the switch to farming. Some see it as a big leap- to a completely different sort of work. Charlie doesn't believe there's such a fundamental difference between the two.

"They both have to do in a broad sense with helping folks- helping all of us to live the good life that God wants us to live."

The strong connection that Charlie feels with the land is obvious, and it's a connection that stems from his childhood.

"Our farm is a spiritual practice" he says, "it's food for our bodies and food for our spirits as well. I feel a real sense of contentment because I've come full circle- I grew up on a farm. I'm doing something that I know was bred into me and it's very satisfying."

Merriam also works on the farm, along with their two youngest children, Katie and Paul. The older siblings, Katherine and Jason, have since moved out on their own. Jason, however, now lives nearby, and he recently helped to build the new barn on his parents' land. As a family the Maloneys raise chickens for eggs, goats for milk, and two hogs which they slaughter each year for meat. Katie keeps a horse as well. It shares a small barn with the goats, and has a pasture on the north side of the property next to the potatoes and salad greens.

*   *   *

Back out in the fields with Steve and Julia, we begin weeding the basil and parsley. We use hoes when we can, though much of the work is again by hand, so as not to endanger the herbs. After we finish we bring bales of straw from the barn to spread around the base of the plants, covering the irrigation hoses. The straw helps prevent the weeds from re-growing, and keeps moisture in the soil.

At noon the lunch bell brings us out of the mid-day sun. We gather around the kitchen and take hands in silence. After a moment a light squeeze of fingers works its way through the circle and it is time to eat. There is salad, bread, maple butter, beans, goat cheese, chicken, and custard pie. Many of the ingredients come right from the surrounding fields. We sit on the screened-porch together, mostly taking a deep breath and cherishing the opportunity to be in the shade.

That afternoon we string some tomato plants to keep them from spreading over the ground, and then move on to pulling heads of garlic. Pulling garlic is the easiest task I would perform during my whole time on the farm. By grabbing the two foot stalk and giving it a swift tug, the garlic pops right out of the soil. That is followed by the favored job of peeling and trimming the garlic, which we do in the cool shade of the barn.

We end the day at around four o'clock. There is plenty of time for a well-deserved outdoor shower before Steve, Julia and I pick some vegetables to go with the falafel we are cooking for dinner. We find tomatoes, zucchini and lettuce for a salad, and cucumber and dill to make a yoghurt sauce for the falafel. Dessert is a berry pie, courtesy of Steve.

It is almost the longest day of the year, and we have filled our stomachs and are playing cards by the time the orange sun dips below the trees on the western side of the farm. When the stars appear I am ready for bed, and I make my way back to the camper. As I lie in my sleeping bag I can hear the creek in the woods behind me, and the crickets in the fields. I can even make out voices that are carrying across the fields from the farmhouse. I fall asleep to the steady beat of the large metal fan that is cooling down the greenhouse, after a hot day in the sun.

*   *   *

At Dayspring Farm, Charlie does not use chemical fertilizers, herbicides, or pesticides, or any inputs that would not comply with Organic standards. He uses strategies such as crop rotation and inter-planting to attract the right types of bugs. Weeds are removed by hand, hoe, or tilling. Natural fertilizers and composts, as well as cover crops, give the soil nutrition. As Charlie and his family attempt to head in a more sustainable direction, they have grants to experiment with no-till agriculture- letting the crops grow in a more natural and diverse environment.

Despite those measures, the farm is not certified as Organic. To Charlie, it's more an issue of cost than anything. While there is no actual fee to be certified, it would realistically cost Dayspring Farm a few thousand dollars in paperwork and labor to go through the entire process. Charlie respects those farmers that do have an Organic certification, but he doesn't feel it's necessary in his situation.

"People do care that we grow the way we grow" says Charlie.

Since he's been able to obtain a local market for his food, he doesn't need to use words like "Organic" for his customers to know that he's being ecologically responsible. They understand the complexities of the certification process and they know why he doesn't bother with it. They know him personally, they know his farm, and they know that the methods he uses are sound.

Charlie tells me that in some ways his farming techniques are similar to those his parents used.

"They did utilize a lot of what we're now calling sustainable agriculture" he says.

On his parents' farm, for instance, most of the inputs came from the farm itself- such as the feed for the various animals.

It was around the time of Charlie's youth that many farmers were starting to experiment more with synthetic chemicals on their land. Charlie says that his father did start to use some chemical fertilizers, but they largely avoided pesticides and herbicides.

"I think he had a sense that it wasn't good. It was real different from the way he grew up farming."

And though he may avoid chemical inputs like his father once did, Charlie doesn't see what he's doing as returning to the past. He uses very intentional methods such as inter-planting and crop rotations, which are things he says his father never learned to do.

"I believe there is a lot we can learn from the past. I value the way farming was traditionally done" says Charlie. "It's going back and learning from the past and picking up some of those practices and combining them with some of the new things we now know and technologies we now have."

Charlie is always learning and trying to improve his methods. In some ways his farm is a constant experiment to see which crops he should plant near each other to ward off pests, or which fertilizers work best at different times.

"This is very challenging, to integrate all the complexities of the farm" he says.

And Charlie must manage those variables for a few hundred plant varieties, and make it all function within his 18 acre environment.

"I've always been a hard worker. I like the challenge."

Charlie's local sales center around a subscription market known as Community Supported Agriculture (CSA). He learned about the concept from reading the book *Master Gardner* by Elliot Coleman. Coleman called the idea a guild, but the theory is the same. CSA's have members, or subscribers, from the surrounding area who commit to buying the farmer's food at the start of each year. As is the case with Dayspring Farm, many CSA's have their subscribers pay the entire season's cost upfront. Each week they pick up their share of food at the farm or some other central location. With almost no cost for retailing, advertising, or shipping, CSA prices are competitive, especially when compared to the premium cost of other organic products.

Most CSA farms follow Organic or even Biodynamic guidelines. They often sell excess food at farmers markets or roadside stands. Charlie has a relationship with four local restaurants who buy from him each week. They constitute the remainder of his market, and also appreciate his farming methods even if he has no official certification.

It's the desire for local, sustainable, and seasonal food that draws in subscribers. They enjoy a grocery bag that's full of surprises, since they don't get the option of choosing what's ripe. And the farmers usually pick the items that morning, or at worst the day before, so the produce is always fresh. The guaranteed market and shared risk both make CSA's ideal for small farmers like Charlie.

The CSA concept started in Japan during the 1960's with local food partnerships called *teikei*. It soon moved to

Europe; there they are frequently called "box-schemes" (since the food each week is delivered in a box). The founding ideas behind these farms were food safety, ecological sustainability, and social equality. In Europe especially, the first CSA-type farmers were following some of the economic and agricultural ideas of Rudolf Steiner.

In 1985 it was a man named Jan Vander Tuin who brought the idea from Europe to the United States. Vander Tuin teamed with Robyn Van En to start the first CSA in North America- the Indian Line Farm in Great Barrington, Massachusetts. In their first year they gave their subscribers apples, vinegar, and cider (some of which was hard). The following year they included a range of produce items, and continued to expand after that.

Robyn Van En traveled the country to promote sustainable agriculture and the CSA concept, and many people credit her with its conception. But there were others involved. The Temple-Wilton Community Farm in New Hampshire was another CSA that began at nearly the same time. From these first seed farms the idea has flourished around the country.

When Charlie started his CSA he had about 20 members. Today, Dayspring farm has grown to include 155 shareholders. About 70% of them order a half-share and the rest order a full-share. In addition, Charlie now offers a flower share and an autumn share. He separates the summer and fall growing seasons because he knows that not all of his members will want to eat the hearty, fall crops such as kale. After he makes the restaurant deliveries each week, he, his family, and the other workers eat the rest.

Charlie feels he could expand his farm even more if he wanted to.

"The market is out there" he says.

In fact he is keeping a waiting list for new members. He may increase his restaurant sales or set up a roadside stand in the future. But for the moment, he is content strengthening the market he has.

"I think we are under-pricing still."

When I ask Charlie if it's been difficult for him to make money as a farmer he chuckles.

"We've been through so much throughout the years" he says.

It seems he has even faced moments of doubt. Now, after making investments in things like greenhouses, the new barn, and irrigation systems, Charlie believes his efforts will start to pay off.

"We've worked really hard to get to this place."

And it's the subscribers that reinforce his success.

"Folks want us to do well, want us to succeed" says Charlie. "Overall, folks are really happy to be getting what we're providing. We're providing good quality food; good quality products. It makes them healthier, makes them feel better, and puts a smile on their face. I find that to be very satisfying. It's very positive work."

*   *   *

It's Tuesday, and it's market day. I start the day picking dill and basil. Soon after, Steve, Julia, Zach and I collect green beans together. We take only the larger pods, leaving the small ones to grow for the weeks ahead. As we work down the rows we also pull up the yellow-flowered mustard plants, intruders, to let them decompose on the ground. Charlie is across the farm picking squash. Once we collect all we need for the day, we divide everything among the subscribers' bags. After another group lunch, we load up the truck.

The destination is Williamsburg, Virginia, about 30 miles to the southeast. At around 3 o'clock we arrive at a house and set up under the shade of the garage roof. The house belongs to one of Charlie's subscribers, and she lets him use her driveway as a pickup station for other members.

This week Charlie is giving his subscribers small red potatoes, green garlic, cucumbers, dill, beets, summer squash, spring onions, "provider" green beans, and long leaf basil. The bags also include a sheet from Charlie with information on the week's take. It talks about how the weather has been, and what that has meant on the farm. It highlights some of the activities they have done that week, such as transplanting nearly 3,000 sweet potato plants. It even includes some nutritional information and cooking tips. "Beets are a good source of folic acid" it reads, and "dill- nice to use with your potatoes and cukes." At the end it is signed "Blessings, The Maloneys," with their home telephone number.

It is Merriam who has driven the truck to Williamsburg on this particular day. Many of the shareholders have been members for years, and some stop and chat with Merriam for 10 or even 20 minutes. A few inquire who I am, as they have never seen me before. When I tell one man I am interested in climate change, he tells me I should work on drawing a map of what Virginia will look like with potential sea level rise. It is of special concern to him since he tells me he lives on the waterfront.

"We'll have to move to the mountains" he says.

CHAPTER 9
Eating Right

Those concerned with the issue of climate change often wonder what they can do on a personal level to help avert what may be a pending global disaster. More and more individuals are changing their light bulbs and using public transportation. One area that people often overlook in their quest for climate neutrality is food. By choosing produce from Dayspring Farm, for instance, Charlie's subscribers are making a connection with the environment that helps to lower greenhouse gas emissions in two main ways.

The first reason, and perhaps the most obvious, is that the food is local. Today in the United States, people are eating more food that travels farther to get to their plate. International food markets are growing faster than any others, and the average prepared meal in the U.S. contains five ingredients from beyond its borders. The farther a bite of food travels, whether it is on a truck, ship, or plane, the more carbon emissions it's tied to.

Researchers at the University of Iowa studied the distance that fresh produce was traveling to reach consumers in their state. Under a system of local growth and distribution, the food was moving an average of 56 miles. Under the normal trade scenario, the food was moving an average of 1,494 miles, almost 27 times as far. The local food had a lot less fuel behind it.

The second reason that Charlie's food is more carbon-friendly is because it's farmed with organic practices. At the Rodale Institute in Pennsylvania, researchers have been studying the differences between organic and conventional farming systems since 1981. Over that time, the energy inputs for organic corn has been 30% less than for conventional corn plots. One of the biggest reasons for the disparity is because the organic corn uses natural instead of industrial fertilizers. Industrial nitrogen fertilizer in particular requires a tremendous amount of energy to produce, and therefore emits more carbon.

Organic farming also creates soils that physically contain more carbon than would conventional techniques. Soils are the largest carbon sink on the planet, holding over twice the amount of carbon as all of the earth's vegetation. Soils also hold twice the amount of carbon than is in the atmosphere. The carbon stored in dirt exists as organic matter from both living and dead plants, insects, and microbes. Much of it escapes to the atmosphere naturally from decomposition and mineralization- but that happens as part of a normal cycle, where the carbon in the soil is also constantly being renewed. When humans disturb the soil with farming, we speed the carbon release and decrease the soil's capacity to retain more of it.

In 1981, the Rodale Institute measured soil carbon as equal in both the organic and conventional plots. When they took measurements again in 2002, the soil carbon in the organic plots was 20-25% higher. By avoiding chemical inputs and nurturing the living organisms in the soil, the organic plot had acted as a much better carbon sink.

Besides eating local, organic meals, consumers can reduce their carbon footprint through food in a third way- by eating less meat. Reasons behind this fact are more complex, but in reality the world's livestock industry is responsible for a large quantity of greenhouse gases.

In a 2006 study titled "Livestock's Long Shadow," scientists working for the United Nations reported that worldwide, livestock is responsible for almost one-fifth of greenhouse gas emissions. That number begins with the livestock industry emitting 9% of the world carbon dioxide. Carbon dioxide (CO_2) is the primary greenhouse gas, simply because humans produce so much of it, but there are others as well. Methane (CH_4), which has 23 times the global warming power of carbon dioxide, and nitrous oxide (N_2O), which has 296 times the global warming power of carbon dioxide, are the two other principle gases. The report attributes 37% of methane emissions, and 65% of nitrous oxide emissions to livestock. In the end, those gases are enough to be responsible for 18% of human induced climate change.

Where do these numbers come from? The sources of livestock's carbon footprint are numerous, and many of them are partly from the farms that grow feed for the animals.

- **Transport of livestock.** This includes moving animals to feed yards and slaughter houses, as well as their shipment to grocery stores: 0.8 million tones[7] of CO_2 per year
- **Land use changes.** Converting forest to cropland releases carbon from the vegetation and soil: 2,400 million tones of CO_2 per year
- **Soil cultivation.** This aides the loss of soil carbon: 28 million tones of CO_2 per year
- **Desertification.** Desertification happens faster under livestock pasture than under any other land use. Hoof action, in an overstressed environment, can turn grassland to desert. This releases much more carbon from the soil and vegetation than with simple cultivation: 100 million tones of CO_2 per year
- **Livestock Processing.** There is not enough data on this for a strong estimate, but meat processing contributes somewhere in the tens of millions of tones of CO_2 per year
- **Farm Energy.** The fossil fuels that the farms use themselves. This covers livestock farms, as well as farms that grow feed for livestock. It can include tractor use, electricity, heating and

---

[7] These figures are in metric tones. 1 metric tone is equivalent to 1000 kilograms. It is also equal to 2,205 US pounds, or 1.1 US tons.

cooling, and the manufacture and processing of seeds, herbicides, and pesticides: 90 million tones of CO_2 per year

- **Manufacture of Fertilizer.** Most of the crops which animals eat every day were farmed with fertilizer. The industrial process to fix nitrogen and make that fertilizer is highly energy intensive: 41 million tones of CO_2 per year

Nitrous oxide emissions, too, mostly come from fertilizer. Plants will absorb only part of the nitrogen that farmers apply to the soil, and the rest will often pollute waterways before breaking down into the most potent of the three major greenhouse gases. Methane emissions primarily come from the digestion of animals. They emit 86 million tones of methane gas each year from the work of their internal organs. Manure emits another 18 million tones of methane per year. The scientists acknowledge that much of the data is difficult to calculate and imprecise, but they also say that the numbers they use are conservative.

It is obvious that if we reduced our meat consumption, greenhouse gases emissions would also go down. But some of the emissions sources, such as transportation or fertilizer use, might appear to relate to food in general, and not specifically with meat. If we replaced meat with grains, vegetables, or other farmed products, would the emissions from things like farm energy, soil cultivation, and fertilizer remain equal?

In fact, the answer to that critical question is no. If we ate less meat we would be eating more farmed products, but we would actually need to farm less. The amount of farmland we devote to feeding livestock is far more than we would need to feed ourselves those extra vegetarian calories. Livestock animals are living, breathing creatures. Feed they consume doesn't just work to create edible meat, it also works to create bones and organs, and to grow hair, and to provide energy to walk, run and breathe.

Because an animal is alive, the pounds of meat that come from a chicken or cow will never equal the pounds of corn or grain that went into their mouths. Poultry requires between 2.1 and 3 pounds of feed to produce one edible pound of meat. Pork requires between 4 and 5.5 pounds of feed. And beef needs 10 pounds of feed for that same, single pound of edible flesh.[8]

Better breeding techniques and better feeds are slowly improving these ratios, but it is a basic fact that they will never be 1:1, or even close to it. From an environmental and energy perspective, meat production is inefficient. With less livestock we would need less farmland- calories from crops would go directly to humans instead of being lost in the food chain.

According to organizations such as the USDA and the UN, crops can provide much more protein per acre than livestock. Rice provides 261 pounds of useable protein per acre; corn provides 211, and other legumes

---

[8] These numbers, not surprisingly, are debated by various groups. The US National Cattlemen's Beef Association claims a ratio on only 4.5 to 1 for beef. The US Department of Agricultural Economic Research Service says the ratio is 16 to 1. The more moderate numbers presented here (10 to 1 for beef) were first published by the US Council for Agricultural Science and Technology (CAST).

provide 192. Livestock gives an average of just 45 pounds of useable protein per acre, and beef a mere 20. Once again this shows that by switching some of our calorie intake from animals to plants, we would need less farmland. By doing so, we would reduce carbon emissions from sources like fertilizer production and soil cultivation.

Unfortunately for the environment, meat consumption is increasing. Already, the livestock sector uses 78% of the agricultural land in the world, either to grow feed crops or for grazing. That area covers 30% of all ice-free land on the planet. As the human population grows, livestock's demands on land use will increase.

Livestock production is growing faster than any other agriculture sector in almost every country, and animals are becoming more and more dependent on inexpensive feed crops as opposed to grazing. Humans converted more land to crops between 1950 and 1980 than during the previous 150 years, and most of that cropland went towards feeding livestock.

Growing meat demand will come from both higher populations and higher incomes. The United Nations projects world population to flatten at 9.5 billion people in 2070. As many countries continue to develop, there will be more people with more money. The trend between higher incomes and increased meat consumption is a well documented one.

There are, however, ways to lessen livestock's environmental burden. More efficient farming methods such as conservation tillage can help preserve soil carbon. A switch to organic farming practices would do the same, as well as reduce the energy demand for industrial fertilizers. Technological advances are driving higher crop yields on equal land areas (although at a slowing rate), reducing the need for deforestation. Better diets for animals lead to less methane emissions. Industrial farms now send manure to large, liquid lagoons, which produce most of the methane emissions. Manure in dry form, used as fertilizer in for crops, produces far fewer greenhouse gases. Researchers are even finding ways to re-use that methane as fuel.

The best way to lower emissions from livestock farming may involve a completely different idea of how livestock should be fed. A growing number of farmers today are rejecting the idea that they should purchase corn and soybeans to feed animals like cattle, and turning to a more traditional, and simple technique- grass. By letting a herd of cattle graze, a farmer is eliminating a number of steps from the way modern farms operate, and therefore making huge reductions in their climate impact.

Much of the greenhouse gases that are tied to livestock come from the farms that are used to feed the animals. By relying on grass, farmers eliminate that entire source. There are no fields of crops that release organic matter from the soil. There are no large machines to harvest the feed. There is no fertilizer that would be applied to the fields which grow that food. There is less transportation of the animal and no transportation of what the animal eats.

Grazing animals are not always the perfect solution. An overgrazed area can lead to desertification and the release of huge amounts of carbon. But a well managed pasture can do the opposite. Many grass-fed herders today practice a style of farming they call 'management intensive grazing.' They continuously move their herd from one small plot to the next, letting the pasture recover for enough time, and ensuring that the cattle eat the grass at the proper growth period to maximize nutrition. Instead of releasing vast amounts of carbon, the pasture stores it. In fact, some studies show that livestock raised on well-managed pasture can even be a net carbon sink. Pasture that is not tilled holds carbon in the soil and vast root structures of the small plants. Grasslands can store even more carbon than forests. Manure from the animals fertilizes the fields, giving it more stability and vitality. No longer a source of methane pollution, the manure actually becomes an asset.

The benefits of grass-fed livestock don't stop with the climate. Pastured meat is touted to be far healthier than meat fed on grain. Grass-fed has both less fat and fewer calories. It has a more ideal ratio of the essential fatty acids, omega-3 and omega-6. Omega-3 helps the blood flow freely, and omega-6 helps it clot. Both are important properties, but too much of one or the other can be harmful. The ratio of omega-6 to omega-3 in grain-fed animals ranges from 5:1 to 14:1, which not surprisingly may lead to strokes and heart attacks. In grass-fed meat the ratio is about 3:1, and sometimes a low as 1:1. Grass-fed meat also has more vitamin E, vitamin A, and other health-promoting substances such as conjugated linoleic acid, folic acid, and beta-carotene.

The livestock industry is of great importance. It is not something we can totally abandon simply because it is helping to warm the climate. Meat provides a critical source of protein for people around the world. In many countries livestock animals are of tremendous social importance. In developing countries especially, livestock provides food security in case of crop shortages. While the livestock sector is only 1.4% of the global economy, it is the livelihood for many poor people who otherwise would have nothing. In wealthy countries, where meat is often a large part of the diet, people could reduce meat consumption with no adverse affects to their health or society, and have great impact towards helping the environment.

Since we eat three meals a day, many of us might consider food our closest connection to the natural world. We have the opportunity each time we eat to decide how we make that connection, and what type of connection it is. If we as consumers decided to eat more local, organic produce- like the food that Charlie provides his shareholders each week- and more grass-fed meat products, we deliberately choose a connection that respects the environment and reduces greenhouse gases emissions.

*   *   *

Whether it's because people are trying to reduce their carbon footprints or not, the demand for local and organic foods and grass-fed meat is on the rise. The local food movement is surging with farmer's markets and buyers clubs sprouting in cities and towns everywhere. The CSA movement in the United States, which started with examples like Robyn Van En's Indian Line Farm, is now enormous. The Robyn Van En center for CSA Resources lists about 1,200 CSA farms in their online database, and that growth shows no sign of slowing down. In similar fashion, author and grass-fed enthusiast Jo Robinson manages an online database of grass-fed farms nationwide. It lists over 800 locations that supply beef, poultry, eggs, dairy, and more. Most of the farms sell locally, and many engage in CSA-type arrangements.

The biggest movement of all is in Organic foods. Today there are hundreds of organizations around the globe that certify farms as Organic. By 2007 almost 634,000 farms worldwide had been officially recognized, covering 76.6 million acres of farmland. In 2005 the Organic market was worth $33 billion- a 43% rise from only three years before.

Dayspring Farm is in King and Queen County of Virginia, an area that Charlie tells me is mostly dominated by agriculture and forestry. Surrounding his land are corn, bean, and barley farms, along with some small dairy herds. While most of those farms use synthetic chemicals, Charlie thinks that organics is starting to catch on.

"I believe that there are increasing numbers of farmers that are curious about it" he says. "They're interested. They want to learn more."

Charlie tells me that there is even one soybean farm nearby that is certified as organic.

"More farmers around recognize that it is an up and coming thing."

Organics is still not as widespread in North America as it is elsewhere, but the continent has been experiencing the fastest growth. North America now has over 12,000 certified Organic farms, and the total acreage increased by 30% between 2004 and 2007. Dayspring Farm might not be certified, but it is still part of the movement. And there are countless other farms like Charlie's that obey organic principles but don't fall under the statistics.

I ask Charlie if he feels that everybody should eat organic food.

"That would be ideal."

He reasons aren't because it might increase soil carbon; he just doesn't believe that synthetic chemicals belong anywhere near the foods we eat.

"We've done a tremendous amount of damage to the farms and to our health with the use of chemicals" he says.

Eating local might be the most important step of all. As Charlie knows, when a farmer sells to a local market they develop a relationship with the buyers; that relationship between farmer and buyer would never happen if the food traveled hundreds or thousands of miles to a grocery store. According to Charlie, it's those relationships that help create better farming practices. Local buyers help promote organic foods or grass-fed meats. And it's the community surrounding those farms that helps sustain them.

"It's important for folks to know where their food is coming from" he says. "The more we eat locally, the more likely we are to encourage responsible production of our food."

And as a farmer that sells locally- "you're more likely to grow with your consumer in mind."

Organic, local, and grass-fed foods are catching on, and there are many reasons for it. It might be for personal health, or to help the local economy. It might be because people don't want to pollute their soils and waterways with synthetic chemicals. Whatever the reason, these foods are also helping to prevent climate change.

CHAPTER 10
Global Food

What climate change will mean for a small farm like Dayspring it's difficult to know. Like at Morse Farm in Vermont and the Benziger ranch in California, Dayspring will certainly face new challenges with heat, pests, weeds, and water. And as with Burr Morse and John Garn, Charlie had already noticed some changes.

"I know some things are happening" he says.

One thing he has observed is warmer winters and a longer growing season.

"I've seen this in my lifetime. We know that colder winters help to take care of some of the pest problems."

And again like at those first two locations, Charlie fears not just the rise in average temperatures, but also the change in extremes.

"The weather just seems more erratic" he says.

As the changing climate starts to take shape, Charlie hopes to find ways to adapt. In recent years he has been adding more irrigation piping. More greenhouse tunnels give him a place to germinate crops in a secure environment. He tells me that in the future he might consider using shade-cloth during the hotter parts of the year.

Adaptation for Charlie could also mean altering his crop rotation and inter-planting strategies to deal with new pest and weed problems. It could mean continuing along a path to more no-till agriculture. It could mean different types of fertilizers. Until the crops, weeds, and pests start to find new niches in a new environment, it is hard to be sure.

Unlike maple syrup or wine production, Charlie may have an advantage in a changing climate. He doesn't depend on just trees or vines- both of which require specific climate conditions to yield their product. Instead, he has many varieties of many different plants. Some crops may suffer in hotter temperatures; others may excel. Certain varieties can be replaced with others, and that can happen much more quickly than the time it would take to grow new trees or vines. But still, changes will involve experimentation in a system that Charlie already describes as complex. How quickly he can realize the changes he needs to make may determine his success more than anything.

*   *   *

Regardless of what happens on any single farm, it's also hard to determine what will happen to food supplies across the entire country. In the United States, climate change may actually have the potential to increase food production. Scientists believe that warmer average temperatures would boost crop growth in many areas of the country- even with relatively high warming by the end of the century. The harvest of staple foods such as corn, rice, soybeans, wheat, and some fruits, could rise by 5-20%- meaning there would be more food available, and it would be cheaper.

However, scientists acknowledge that those projections do not take well enough into account the rising extreme temperatures that will come with rising averages. While crops could benefit from higher averages, they could be equally hurt by more intense heat waves and other short-term variability. When average temperatures rise, so does the entire range. What was once an extreme event becomes more normal, and the new extremes move into unknown territory.

Even if the country sees overall gains in food production, that would not tell the whole story. With just mild or moderate warming, some crops, and some whole regions of the country will face declines. Maple syrup and wine production are only two examples. In the plains, dry-land wheat crops might decline by 10-50%. In the southeast, soybeans- a major crop there- could face yield reductions of up to 70%. Rice and tomatoes would also decline in the southeast, while citrus production would rise because of less cold-weather damage.

Farmers everywhere could face additional problems besides those related to average temperatures or even extremes. With less snowpack and more soil evaporation, water shortages could be a problem. This is especially true if some areas start to receive either less rain, or rain at different times. Heavy rains during shorter periods would cause topsoil erosion. Especially in the west, farmers could face a greater risk of wildfires. Warmer winters would help create additional challenges with pests and weeds. More carbon in the atmosphere would mean faster plant growth, which would also mean a higher susceptibility to pests and a greater need for nitrogen fertilizer. Scientists predict that with climate change, nitrogen-loading into the Chesapeake Bay would increase by 17-31% from farming applications. In a body of water on that is already severely polluted, that will not be a welcome change.

To take an optimistic view, the United States will be capable of adapting to certain setbacks. Some areas may face losses, but other areas of the country could make up for them. Soybean farmers in the southeast, for instance, would lose a competitive advantage, and production would shift to more suitable regions. For more specialized crops like maple syrup, we are already seeing that Canada is primed to fill demand if places like Vermont see large declines. Perhaps we would find other places to grow high quality wine-grapes.

But shifting production doesn't change the fact that there will be losers. For generations we will find farmers who either need to relocate their home and business or find a new line of work.

Shifting or expanding farm production also causes other problems. By moving such a major land user to new territory we takeover more wild land- causing the further destruction and fragmentation natural habitats. Humans can turn a forest into a corn field much faster than nature can turn an old cornfield in a forest. This would mean accelerated losses of biodiversity and ecosystem services. Relocating farms also puts them at odds with human developments, further tightening the land squeeze in a country that continues to grow and expand.

*   *   *

On a global scale, the future of farming is similar to that of the United States. With moderate warming- in the range of 2-5° Fahrenheit- mixed with some shifts in rainfall and an increase in atmospheric carbon, scientists predict that total food production could rise. The total land area available to farming would increase, and food prices would fall.

But also like in the United States, those predictions are based on average temperatures, and they face many uncertainties involving extremes. Extreme temperature ranges have the potential to overcome those gains in crop yields, and push them into declines.

Extreme events stem from both new ranges and greater unpredictability. Heat is energy. As the climate warms it puts more power into the weather system. This is the reason why we could see more storms and erratic patterns. Temperatures will be affected along with winds and rains. Without knowing what to expect, it can be hard to prepare.

As one example of what extremes can do to agriculture, a heat wave devastated crops across Europe in the summer of 2003. In some places temperatures rose to 10° F above the norm, and precipitation declined by almost 12 inches. In the Po Valley in Italy, the corn harvest fell by 36%- a record loss. In France, corn was 30% below the 2002 harvest, and fruit was down 25%. It was the continent's worst year for wine production within the previous ten. In the end, Europe endured uninsured losses of €13 billion.

If average temperatures alone rise past the moderate range, global food supply will likely begin to fall. Scientists predict that with a rise of 10°F, food production will fall by enough to cause global prices to rise by 30%. And again, those changes fail to account for extremes.

Whether we see overall gains or declines, many areas will experience regional losses- much like in the United States. With only mild warming, scientists predict crop losses in low latitude regions. This is especially true in dryland tropical areas such as sub-Saharan Africa and southern Asia. Those regions include mostly developing countries that already face many problems, food security being one of them. By 2080, researches predict that 75% of all people at risk for hunger will live in sub-Saharan Africa. Climate change is not the only reason for this disparity, but in the coming decades it will be one of the driving forces.

A great number of people that live in low-latitude developing countries are small-scale and subsistence farmers. They live in rural settings; agriculture is their major source of income and livelihood and they rely mostly on family labor. In fact, 70% of the people in developing countries still live in rural settings- where agriculture dominates. Small-scale farmers face significant threats to food security, especially when climate change becomes a factor. If climate change doesn't cause crop shortages in those areas by itself, it has the potential to merge with other environmental and social stresses to achieve the same result.

Median United Nation projections place the world's population at 9.5 billion by 2070. 95% of that growth will happen in developing countries. To feed these people, researchers expect farmers to takeover another 605 million acres of cropland. And again, this will happen almost exclusively in developing countries. Farmers will also need to increase yields from their current rate of 2.7 tones per hectare to 3.8 tones per hectare by 2050.[*] If these improvements happen, and other factors do not diminish food supplies, crop production would rise by 80% by 2050, and the number of undernourished people would decline.

Yet in the developing world and elsewhere, people face the dual challenge of food demand and environmental protection. Soil erosion, salination of irrigated areas, groundwater loss, chemical pollution, and nutrient loss are all becoming serious issues for agriculture. Even as more lands in developing countries become marginal for food production, farmers may cultivate them anyway under growing land pressure. This could lead to deforestation, biodiversity loss and desertification. Today, deforestation

---

[*] 1 hectare is equivalent to 2.47 US acres.

continues at a rate of 13 million hectares per year. In the end, the loss of ecosystem services could threaten food security even more in the face of climate change.

It seems almost certain that under warming conditions some high-risk areas will face losses. Fortunately, we have the potential to offset those losses with trade. But trade implies that the poorer countries will need to buy more food. A large population of people in the developing world base their incomes and livelihood on agriculture. If those farms become unmanageable in hotter, drier weather, where will they find the money to pay for more food? If there are food shortages caused by more advanced warming, prices will rise. At the same time, poorer countries will need to increase their cereal imports by up to 40%. Even today, the global food supply is more than adequate to feed the planet's population, yet there are 800 million undernourished individuals. If the poorer countries lose the ability to produce much of their own food, international trade won't fix the problem alone.

Of course the picture is not all negative. Because much of the developing world still relies on less sophisticated farming methods, there is more room for improvements and adaptation. Farmers there could learn many new techniques involving crop varieties, planting and harvest times, fertilizer application, efficient water use, and pest, disease, and weed management. But turning these adaptations into reality is not always easy. Small-scale farmers will need assistance in the form of aid money, technology transfer, and education. As climate change becomes more pressing, farmers like Charlie Maloney may find ways to adapt to hotter temperatures and extreme events, but those without resources and infrastructure will not. Hopefully there will at least be opportunities for the world's small-scale farmers to adapt, though we can expect that not all will succeed. When temperatures rise higher, adaptation will become even more difficult. Rich countries should aid the poor not just because they will have the resources to do so, but because they are the ones who put the vast majority of greenhouse gases into the atmosphere to create the problem. Millions, perhaps billions, will need to find new ways to sustain themselves, and that will require a commitment from the world as a whole.

CHAPTER 11
The Future of Food

Modern farming is a constant battle against nature. The primary tools of the industrial farmer—pesticides, herbicides and fungicides—stave off the crop's enemies, but at the same time they strip the soil of vital bacteria and nutrients. Farmers must replace those nutrients with greater and greater quantities of chemical fertilizers. They destroy predator-prey relationships. They halt genetic variance and make evolution the work of laboratories.

Meanwhile, some food producers choose a very different path. Farmers like Charlie Maloney and Mike Benziger build on the wisdom of past generations and use new ideas to move into the future, still keeping the idea of sustainability in focus. In order to nourish their crops they work *with* the natural systems of their farms, they don't suppress them.

How did we get here? And how did the act of fueling our bodies become so complicated? Should we accept industrial agriculture as the norm, while words like 'Organic' are footnotes on a few supermarket shelves? Food is too important to ignore, and climate change may threaten our supply. We must understand food production today, and decide which pathway to take from here.

*   *   *

Archeology can tell us where and when agriculture began. The Neolithic revolution started at around 10,000 BCE in southwestern Asia, in the "fertile crescent" of ancient Mesopotamia- modern-day Iraq. Farming also developed soon after in other parts of the world, such as China, America, Africa and Southeast Asia. Yet the most critical question might be why did agriculture begin?

Agriculture is among the most important developments in human history; it gave birth to cities, and cities gave birth to modern civilization. But why did humans make the switch to farming? One prominent theory is based on something that is happening again today- climate change.

The majority of our ancestors were hunter-gatherers, and for tens of thousands of years they were content with that lifestyle. Some people assume humans started farming when they first learned how. Most historians, however, don't believe that to be true. The Neolithic revolution was not a sought-after change.

To a group of people that could find and eat wild foods, planting crops or feeding animals was an extra step. Nature did that for them. Hunter-gatherers were adept at their lifestyle, and they found their food with skill. Life wasn't posh, but it was simple, and they spent most of their time in leisure. When they did start to farm, the workload became far more demanding. And they had to make social changes- adapting to life with larger groups of people. For generations our ancestors probably understood the concept of planting seeds, but chose not to. When they finally did, there were forces outside of their control.

It seems more than coincidence that this momentous shift in history was mirrored by another major change. At around 13,000 BCE the Pleistocene- the climatic period of the last major ice age- started to end, and the Holocene- our current climatic period- was beginning. Warmer temperatures led to more rain, and this produced more wild cereals and grazing animals. People wouldn't ignore the opportunity to utilize these food sources. Human populations increased, and they began to rely on certain plants and animals.

At around 10,000 BCE there was another, smaller climate period called the Younger Dryas. For about 1,000 years, the planet reverted back to near ice-age conditions. During this harsher epoch there were both more people to feed and less land to feed them. Populations that grew at the start of the Holocene had to join together and intensify food production. Agriculture may have saved them from mass-starvation.

As the climate went through shifts, plants and animals looked for new ways to adapt and survive. Some species died off, and others emerged. In this period of flux it was much easier for humans to domesticate certain foods. But farming wasn't just about dominating the landscape. From the perspective of the plants and animals, they took advantage of humans. In farmers and pastoralists they found protection, feed, and regeneration; except for a controlled death, it was life made easy. Agriculture wasn't only domestication, it was mutual dependence.

*   *   *

To most farmers today, codependence is a thing of the past. Industrial farms try to control every part of their environment. They bring natural evolution to a halt and rely on various chemicals to make advancements. The most perverted path in industrial farming is the push towards uniformity.

With uniformity comes vulnerability. The threats plaguing our crops have high populations and short life cycles. When hundreds of acres of a single crop are laid out for them, it shouldn't be surprising that pests can evolve beyond even the poisons set out to stop them. If we rely on the laboratory, we are playing a never ending evolutionary game to keep them at bay.

With pesticides, for instance, 99% of the chemicals don't ever reach their intended target. Since 1950, insecticide use has risen eight-fold, while at the same time crop losses have almost doubled. And regretfully, we continue to learn more about the dangers of these practices. Pesticides and herbicides are raising serious public health concerns. Studies are connecting them to immune system damage, various cancers, and more.

In livestock husbandry the game is no different. Except here we rely on antibiotics and vaccines to keep the crowded and diseased animals from dying before they reach slaughter.

The genetic base of our crops is shrinking rapidly. Native varieties are constantly wiped out for industrial uniformity. By the early 1990's a mere six varieties of corn made up 46% of the harvest in the United States. Over half of the world's potato crop is one type- the Russet Burbank (the type preferred by McDonalds). Having little genetic variance, plants lack the capability to build up natural resistance.

The most perverted example of genetic uniformity might be the banana. Virtually all of the bananas we eat are a single variety- the Cavendish. In fact the bananas aren't just the same variety; they are all the same banana. To plant a banana tree you don't need seeds, you simply need a graft of another tree. The bananas are clones of each other from grafting tree after tree. Most banana plantations around the world have collectively zero genetic diversity.

Today this army of banana clones has created a serious problem. Bananas are being attacked by a fungus called Panama's disease, and not even scientists can find a cure. Banana growers chose the Cavendish variety because it was supposedly resistant to Panama's disease. The fungus wiped out plantations of the popular Gros Michel banana variety in the 1950's, and Cavendish was the cure. But Cavendish bananas have encountered a new strain of Panama's disease in Asia they cannot resist. Now Panama's disease is destroying plantations across Asia and Australia.

Sooner or later it will make the jump the Americas in some small patch of polluted soil, and it will spread quickly.

Whether it is bananas, potatoes, or any other crop, these industrial monocultures can't adapt well to change. And just like when agriculture first began, we are again experiencing a shifting climate. A changing environment means that pests and diseases will seek new ways to survive. Without genetic diversity or natural resistance, we will have to rely on the laboratory to save our crops. Farmers will apply newer and newer chemicals, and more and more of them. Hopefully they can keep up.

Agriculture did not come about over the course of a few generations. It took millennia for any society to switch entirely to farming. Evolution is slow, and plants and animals needed time to adapt to farmers' needs. Many of the same crops that humans first cultivated such as wheat, barley, rye, lentil, pea, and chickpea, are still important to society today. Without another major climate shift since the start of agriculture, those crops have had no reason to change. With a climate now on the brink, that cannot remain true forever.

*   *   *

What has changed during the Neolithic age are farming practices. For a long time developments were slow. Humans gradually learned to implement the plow, animal power, crop rotations, fallows, manure, and irrigation. They took advantage of more crop varieties, including fruits and vegetables. They learned the best times of the year to plant and harvest. With selection, the varieties improved and yields increased.

In recent centuries these improvements accelerated. The horse replaced the oxen, and then the tractor replaced the horse. Farmers used more external inputs- they learned to fix soil deficiencies with industry. In the 1840's researchers in England developed "superphosphates" in order to put phosphorous back into the soil. The process involved treating phosphate rocks with sulfuric acid, and it was the first artificial fertilizer.

The big breakthrough was nitrogen fertilizer. Around the start of the 20th century Fritz Haber and Carl Bosch developed and commercialized a process to fix nitrogen. By the end of World War II the Haber-Bosch process made artificial nitrogen fertilizer available to all. Chemical herbicides replaced the hoe, and pesticides gave the crops another defense against predators. Irrigation improved. The tractor went from 10 horsepower to over 400. The corn combine replaced the husking hook. Before long, food production exploded.

In 1900 it took a farmer 81 minutes to produce one bushel of corn. 100 years later he could complete the task in just two minutes. During the same period the average corn harvest in the United States rose from 21 bushels per acre to 121. That also made food cheaper. Americans in 1900 spent around 25% of their income on food. By the year 2000 that dropped to 14%.

Of course other advances during the century helped make the agricultural revolution possible. They include the development and spread of cars, plumbing, electricity, sanitation, equal rights, education, and business management. With these structures in place, farmers could concentrate more on their fields and perform better in a stronger society. For all of these reasons, farming changed more in the 20th century than during the thousands of years preceding it.

As we head farther into the 21st century, industrial agriculture is moving down a pathway of more technological solutions. Chemical inputs and machinery, which made the 20th century's food revolution possible, continue to be a major force in agriculture. Now they are joined by even newer advances such as biotechnology.

Scientists have isolated crop genes that control heat, drought, pest, and disease resistance. Altering genes can help crops adapt to more salinity in the soil or carbon in the atmosphere; it can change their flowering and ripening times. Researchers have had some successes. Often, however, plants will compensate for what scientists do in the laboratory and show no real change. How exactly biotechnology will play out in the fields, and how it will endure as a long-term solution, we don't yet know. Biotechnology demands a great deal more research- both of its effectiveness, and of how it impacts human and environmental health. If we want biotechnology to succeed, than we aspire to nothing less than being smarter and faster than evolution, something nature has been practicing for billions of years.

Down another pathway, Charlie, Mike, and other growers are showing that small, diverse farms can be both profitable and sustainable. But, can organic systems feed the world? It was chemical inputs that helped make the 20th century food revolution possible. If we abandon them, could farms keep up with the growing demand?

A number of places around the world investigate how organic yields compare with conventional ones. In some studies the organic systems did produce significantly less food. European tests of wheat, for instance, showed organic fields producing 30-50% less per acre. Scientists believe that much of that disparity was due to a nitrogen shortage in the organic crops. Some will argue that without industrial nitrogen fertilizers, we will never feed a planet of 9.5 billion people, or even the 6.5 billion people we have today.

Yet other studies suggest that organic yields can be just as high as conventional systems. Started in 1981, the Rodale Institute's Farming Systems Trial is the longest running study of Organic versus conventional farming techniques. In Pennsylvania, the Rodale Institute's researchers maintained high levels of soil nitrogen in the organic plots with animal manure and legume crop rotations alone.[*] In the first five years of their tests, the organic corn plots did produce less than the conventional plots; but after that transition period, both produced equally. During drought years the organic plots produced even higher amounts than the conventional ones.

*   *   *

---

[*] Rhizobium bacteria that live on the roots of peas, beans, and other legumes take nitrogen from the air and place it in the soil, converting it to a form that plants can use.

Modern industrial agriculture creates a long list of problems. Today's large-scale farms pollute the air and waterways. They run up perverse government subsidies. They exploit cheap labor, and they force out small-scale farmers. They even degrade the thing they rely on most, the topsoil. A healthy layer of topsoil requires a geologic event to rebuild; it is not renewable on the human timescale. Until grinding glaciers re-emerge or a continental uplift takes place, what we lose is gone for good. The world continues to lose 25 million acres of good cropland per year because of irresponsible farming practices. While these challenges to industrial agriculture pile up, climate change will only create a more urgent threat.

As Charlie Maloney says, organic strategies have advanced greatly over past decades, even without chemical solutions. Abandoning the pesticides and fertilizers of the 20th century wouldn't send us back to production levels of the 1800's. If more farmers today took a serious look at organic strategies, we might find that they are just as productive. And since Organic and Biodynamic farms are designed to work with the natural environment rather than dominate it, they could even outperform conventional systems as conditions change. Well-designed polycultures can produce high yields while at the same time reducing the need for chemicals. Mimicking natural systems, they fight pests and disease with complexity and a larger genetic base. During a time of transition, do we want to be fighting against nature or working with it?

*   *   *

While climate change was present when agriculture began, but since then humans and their crops have lived through a stable environment for thousands of years. If the climate goes through another shift, we know that plants and animals will search for new relationships, and they will require new ways to survive. Some will succeed better than others. As those changes take place, will we adapt along with them? Will we continue to enjoy things like maple syrup and wine, and will we have the foods we need to live and be healthy? Our ancient ancestors may have started farming because the climate forced it upon them. We don't want nature to push us into another decision of that magnitude. It's not that humans can't flourish in a climate different from the one we inhabit today, but it's the transition period that we should fear.

If we reduce our emissions, we won't have to adapt as much or as quickly. Prevention might mean the difference between gains and losses in global food supply. How can food production play a role? Farmers can grow organic foods and raise grass-fed meats. They can market locally. They can be more efficient and use alternative energies. It's true that an industry like wine or maple syrup doesn't have a huge carbon footprint; but there are thousands of such industries, and they all have an impact. Those with the most to lose should have the most incentive to be leaders. Individuals can consciously select the foods they eat- a decision we all make many times each day.

We have been spewing gases into the atmosphere for centuries. Some changes are coming, and we must be able to adapt. Does the answer lie in chemical and artificial means, or with Biodynamic and Organic strategies? Will farmers work with nature's systems or try to suppress them? Will they try to prevent problems beforehand or rely on technology to fix a crisis that is already happening? Whatever direction we choose to take, our solutions must be able to work on both a warmer planet, and one that is in flux. We can only push an overstressed system so far.

Climate change will affect our lives in countless ways. It will create some new problems, and it will bring some old problems to light. How we treat the climate is a single part of how we treat the environment as a whole. Without climate change, coal-fired power plants would still harm our lungs and emit mercury. Nitrogen fertilizers would still pollute our waterways. When the Neolithic revolution was underway, humans needed social changes to live in a more close-knit, stationary society. Ironically, industrial food production with its far-removed consumers may be one factor that now pulls our society apart. If the climate changes again as it did before, will there be more social changes? And will they involve coming back together to treat the environment in a new way? By joining community food partnerships, for instance, we build a stronger connection with the food, the farmer, and the land around us. We see our own reliance on nature; we understand why we must act and be responsible.

A new climate will change our food production and our society as a whole. The world around us could completely transform, and the way we treat the world could transform as well. We must choose to guide food production, and our entire planet, in a more sustainable direction.

AFTERWORD
The Story of a Vegetarian

One of my best friends in college was a vegan.[11] We all used to give her hell for it. Our favorite pastime was to wave bacon or other pieces of red meat in front of her, taunting. She was never deterred, and she talked about all of the wonderful and tasty things that even vegans can eat. That didn't really pull much weight with my friends and me.

"We can eat everything you can eat, and more," we would say.

Of course my vegan friend *could* have eaten those meat and dairy products, she just chose not to.

There are many reasons to be vegetarian or vegan. Hers was the environment. And she wasn't just concerned

---

[11] Vegans refrain from eating all types of animal products, not just meat. That includes dairy, eggs, fish, etc. Many vegans will also eliminate leather clothing and other animal products from their lives.

with climate change; she knew that eating meat was inefficient and an extra burden on the planet. She knew that the food used to feed a cow could instead be used to feed perhaps ten other people. And if not, she knew that the land used to grow feed could instead remain wild. I understood my friend's reasoning, and I agreed with her. But giving up meat just didn't seem worth it at the time.

Was my vegan friend making a real difference? Was her personal sacrifice worth it? She probably wasn't changing the world. In the United States alone, she was only one of 300 million. Most of the rest of those people were, and are, omnivores. Her effect was minimal.

It's true that on a major scale, if we want to solve the problems that overwhelm the planet we need action on a more grandiose scale. The government needs to reevaluate subsidies that support large-scale, destructive farming. Industry needs to wean itself off of the carbon economy. Scientists must continue to unveil the dangerous effects of herbicides and pesticides. We elect leaders for a reason. Most of us hope that they will do their job by protecting us and our world, now and into the future.

If we educate ourselves and look to the current state of affairs, we can see that is not always the case. Many of the paths we are following are not wise ones. Something must be done. Or rather, we must do something. We need to be headed in a more sustainable direction, and fast. We need advocacy. We need activism. We need to keep fighting for changes that should be formed into law. Even as we fight those battles, we should look to ourselves for personal and more immediate change.

Before the Civil War many abolitionists in the northern United States refused to buy cane sugar because it was produced by slaves. They bought maple sugar instead. Again, their individual actions didn't have a huge effect, and can sugar plantations continued to flourish. Like my friend in college, they probably knew that their actions wouldn't stop cane sugar, or slavery, but it's certain that many individuals still maintained their position. They had a moral objection to the product, and they made a conscious choice. Even if the slave plantations continued, it wasn't something that they would help support.

While working in the White Mountains of New Hampshire one summer I met more friends who were vegan. They had similar reasons; they were doing it because they wanted to help the environment on a personal level. One day it all clicked for me. On June 21, 2006, I declared myself a vegan. The very next day I ate a peanut butter cookie knowing that it didn't have real butter in it. It was delicious. Later, my vegan colleague reminded me that it probably had eggs in it. And I found out it had. That night I ate some soup before I realized the meat hidden within. I decided that being a vegan was a big job, and that I wasn't yet up to the challenge. Instead I maintained a role just as a vegetarian.

Is my personal effort making any greater changes? Again, perhaps not, but I still feel a moral obligation to live a life that is true to my beliefs and convictions. I also make an effort to buy more local and organic foods. Even when they are more expensive, I know than I am saving a higher cost that would be passed on to the environment and society.

Large-scale changes may come, and they may not. Either way, I don't want to support those detrimental ways, regardless of the greater effect I am having.

And perhaps I am helping to make real changes. Every time I take a bite of food I am making a conscious choice about what I put into my mouth. Each time I refuse a piece of meat I think about why. As we make these small, personal changes, we remind ourselves what we value, and where our morals lie. Our convictions grow even stronger, and that leads us to push for the bigger changes that need to happen as well. We also affect others around us. When we share a meal we pass our ideas and thoughts on to friends and family. It was the veganism in others that led to change in me. I hope that I can pass the change along to somebody else, too.

We are a society made up of individuals. We can't expect our leaders to bring change if we are not willing to change ourselves. We must show a commitment and become the change we wish to see. If we are true to our own conscience first, then hopefully our leaders will join us in those efforts. Acting out change doesn't mean that we all have to be vegetarians or vegans. It doesn't mean that everything we buy must be local or organic. Even as I call myself a vegetarian, I occasionally eat grass-fed and wild meats when I have the opportunity. There is no single answer.

We eat many times each day. It is not something we can avoid. It is both our most consumptive habit, and our closest connection to the earth- the only home we have. Choosing what we eat is a great privilege, and it gives us great power. We should all listen to our own conscience and search for the right path to take. Preferably, that path will help lead us to a more sustainable future.

Malcolm S. Lewis
April 2008

BIBLIOGRAPHY and FURTHER READING

INTRODUCTION

IPCC. *Climate Change 2007: Synthesis Report. Contribution of Working Groups I, II and III to the Fourth Assessment Report of the Intergovernmental Panel on Climate Change* [Core Writing Team, Pachauri, R.K and Reisinger, A. (eds.)]. IPCC: Geneva, Switzerland, 2007.
The full report, as well a summaries, are available online at http://www.ipcc.ch/ipccreports/ar4-syr.htm

PART 1
Texts

Allen, Brian T. *Sugaring Off: The Maple Sugar Paintings of Eastman Johnson.* Williamstown, MA: Sterling and Francine Clark Art Institute, Distributed by Yale University Press, 2004.
Contains some facts on the history of maple sugaring, particularly in colonial times.

Deerr, Noël. *History of Sugar.* London: Chapman and Hall, 1949.
Cites the use of slaves in early southern cane sugar production in the United States.

Eagleson, Janet, and Rosemary Hasner. *The Maple Syrup Book*. Ontario, Canada: The Boston Mills Press, 2006.
Contains lots of information on maple syrup. It covers the history of maple sugaring, including the Native American legends, as well as production and facts from today.

Hyde, Bruce. *The Travel and Tourism Industry in Vermont: A Benchmark Study of the Economic Impact of Visitor Expenditures on the Vermont Economy- 2005*. Vermont Department of Tourism and Marketing, and Economic and Policy Resources: Montpelier, Vermont, and Williston, Vermont, 2005.
Available online at
http://www.vermontpartners.org/pdf/TourismImpactStudy2005.pdf
Contains information on the impact of tourism for the Vermont Economy.

Kershner, Bruce, and Robert T. Leverett. *The Sierra Club Guide to the Ancient Forests of the Northeast*. San Francisco: Sierra Club Books, 2004.
Contains information on maple trees, notably the record maples in size and age.

Lockhart, Betty Ann C. *The Maple Sugaring Story Book*. Charlotte, VT: Perceptions, 2004.
Contains information on maple sugaring history and the process today.

Morse, Burr. *Sweet Days and Beyond: The Morse Family, Eight Generations of Maple Sugaring*. Poultney, Vermont: Historical Pages Company, 2005.
Burr's book. Contains information on everything at the Morse Farm, from the first generation until today.

State of New Hampshire. *Selecting Trees for Urban Landscape Ecosystems: Hardy Species for Northern New England Communities*. Concord, NH: Department of Resources and Economic Development, Division of Forests and Lands, 1994.
Contains some general information on the maple tree.

Union of Concerned Scientists. *Climate Change in the U.S. Northeast: A report of the Northeast Climate Impacts Assessment*. Cambridge, MA: UCS Publications, 2006. Available online at
http://www.northeastclimateimpacts.org
Contains information on past and future climate changes in the Northeast United States

Web Resources

Information relating to Governor Jim Douglas' actions on Climate Change can be found on the Governor's website
http://governor.vermont.gov/

Information on emissions statistics by country can be found at the website for the United Nations Statistics Division.
http://unstats.un.org/unsd/environment/air_co2_emissions.htm

Information on emissions statistics by state can be found at the website for the Environmental Protection Agency.
http://www.epa.gov/climatechange/emissions/downloads/CO2FFC_2004.pdf

The website of the State of Vermont's Department Forests, Parks and Recreation. Contains information on the Maple Tree, Maple Sugaring, and the history of both.
http://www.mapleinfo.org/

The website of the New England Maple Museum. Contains information on the museum, as well as information on maple sugaring and the industry. The museum itself offers ample information on maple sugaring, particularly in reference to the history of the trade.
http://www.maplemuseum.com/

The website of Morse Farm Maple Sugarworks. Contains information on the farm and maple history, contact information, and a full range of maple products.
http://www.morsefarm.com

Information on state statistics of maple syrup production can be found on the website of the United States Department of Agriculture
http://www.nass.usda.gov/nh/mapleconf2005.pdf

The website of the Proctor Maple Research Center, Contains information on the center, and the total economic impact of the maple industry on the state of Vermont
http://www.uvm.edu/~pmrc/

The website of the Vermont Maple Sugar Makers' Association and the Vermont Maple Foundation, with information on both groups, and Vermont maple syrup.
http://www.vermontmaple.org/

The website of the Vermont Public Interest Research Group (VPIRG). They are the largest nonprofit consumer and environmental advocacy group in the state of Vermont.
http://www.vpirg.org/

The website of the Vermont Attractions Association
http://www.vtattractions.org/

The website of the Vermont Maple Festival, held each year in St. Albans, VT.
http://www.vtmaplefestival.org/

Interviews

Morse Jr., Harry I. "Burr". Montpelier, VT, May 31, 2007.

Perkins, Timothy. Underhill, VT, May 31, 2007.

PART 2
Texts

Cayan, Dan, Ed Maurer, Mike Dettinger, Mary Tyree,
Katharine Hayhoe, Celine Bonfils, Phil Duffy, and
Ben Santer. *Climate Scenarios for California*. California
Climate Change Convention Center, March, 2006.
Available online at
http://www.climatechange.ca.gov/research/climate
/pdfs/CEC-500-2005-203-SF.pdf
Contains information on future climate scenarios
for the state of California.

Jan de Blij, Harm. *Wine: A Geographic Appreciation*. New
Jersey: Rowman and Allanheld Publishers, 1983.
Contains information on the history of wine.

McCarthy, Ed, and Mary Ewing-Mulligan. *Wine For
Dummies*, 4th ed. Indianapolis, Indiana: Wiley
Publishing, Inc., 2006.
A basic overview of wine.

MKF. *MKF Research Report on Economic Impact of California Wine 2006, Updated January, 2007*. St. Helena, CA: MKF Research, 2006.
Available for purchase at
http://www.mkf.com/orderPublications.html
A report by MKF Research detailing the economic impact of the California wine industry on the state of California and the nation.

Stovall, Pamela. *Guide to American Vineyards: A Guide to the Best Wineries for Touring and Tasting*. Old Saybrook, CT: The Globe Pequot Press, 1992.
Contains information on the history of wine in the United States.

White, M. A., N. S. Diffenbaugh, G. V. Jones, J. S. Pal, and F. Giorgi. *Extreme Heat Reduces and Shifts United States Premium Wine Production in the 21st Century*. The Proceedings of the National Academy of Sciences 103 (30), July, 2006.
Available online at
http://www.pnas.org/cgi/content/full/103/30/11217
A report on how climate changes could affect wine production in the United States.

Web Resources

The website of the Demeter Association, Inc., the United States branch.
http://demeter-usa.org/

The website of the Association of Waldorf Schools of North America, the style of schooling developed by Rudolf Steiner.
http://www.awsna.org/

The website of the Benziger Family Winery and the website for their high-end Tribute wines. Both contain information on the winery, its history, and Biodynamic farming practices.
http://www.benziger.com/
http://www.benziger.com/tribute/

A website on Austrian philosopher, and the inventor of Biodynamic farming, Rudolf Steiner.
http://www.rudolfsteinerweb.com/

The website of the California Sustainable Winegrowing Alliance, and the Sustainable Winegrowing Program. Contains information on both, including the workbook that California winegrowers can use to self-assess their vineyard for sustainability.
http://www.sustainablewinegrowing.org/aboutswp.php

The website of the Unified Wine and Grape Symposium, held each year in Sacramento, California.
http://www.unifiedsymposium.org/

The website for John Garn's environmental consulting company, ViewCraft.
http://www.viewcraft.com/

The website of the Wine Institute. Contains information on the wine industry in California.
http://www.wineinstitute.org/

The website of the Professional Friends of Wine. Contains lots of basic information on wine, including facts on the Cabernet Sauvignon grape.
http://www.winepros.org/

Interviews

Garn, John. Glen Ellen, CA, September 5, 2007.

Sound Recordings

Jones, Gregory, Ph.D. *Climate Change.* Sacramento
    Convention Center, Sacramento,
    CA: Unified Wine and Grape Symposium, January
    24, 2007.
    Available for purchase online at
    http://www.allstartapes.com/conferences/conferen
    ce_1104.shtml

Wilkinson, Robert, and Trumble, John. *Cultural Practices to
    Address Climate Change.* Sacramento Convention
    Center, Sacramento, CA: Unified Wine and Grape
    Symposium, January 24, 2007.
    Available for purchase online at
    http://www.allstartapes.com/conferences/conferen
    ce_1104.shtml

PART 3
Texts

Jackson, Wes. "The Agrarian Mind: Mere Nostalgia or a Practical Necessity?" In *The Essential Agrarian Reader: The Future of Culture, Community, and the Land*, edited by Norman Wirzba, 140-153. Washington, D.C.: Shoemaker and Hoard, 2004.
This is a good chapter of a good book. The book itself was created as somewhat of a follow-up to Wendell Berry's classic text *The Unsettling of America*. The Jackson chapter contains some good information on the broad scope of the harms of industrial agriculture. This includes some of the information on pesticide and herbicide use and topsoil loss.

Bell, Michael Mayerfeld. *An Invitation to Environmental Sociology, 2ⁿᵈ ed.* Thousand Oaks, CA: Pine Forge Press, 2004.
An excellent text on the subject of environmental sociology, as well as excellent and thoughtful reading for the concerned environmentalist. Contains information on CSA's.

Easterling, W.E., P.K. Aggarwal, P. Batima, K.M. Brander, L. Erda, S.M. Howden, A. Kirilenko, J. Morton, J.-F. Soussana, J. Schmidhuber and F.N. Tubiello. "Food, fibre and forest products," in *Climate Change 2007: Impacts, Adaptation and Vulnerability. Contribution of Working Group II to the Fourth Assessment Report of the Intergovernmental Panel on Climate Change*, edited by M.L. Parry, O.F. Canziani, J.P. Palutikof, P.J. van der Linden and C.E. Hanson, 273-313. Cambridge, UK: Cambridge University Press, 2007. Available online at http://www.ipcc.ch/pdf/assessment-report/ar4/wg2/ar4-wg2-chapter5.pdf Contains information on how climate change could affect food production on a global scale.

Field, C.B., L.D. Mortsch,, M. Brklacich, D.L. Forbes, P. Kovacs, J.A. Patz, S.W. Running and M.J. Scott. "North America," in *Climate Change 2007: Impacts, Adaptation and Vulnerability. Contribution of Working Group II to the Fourth Assessment Report of the Intergovernmental Panel on Climate Change*, edited by M.L. Parry, O.F. Canziani, J.P. Palutikof, P.J. van der Linden and C.E. Hanson, 617-652. Cambridge, UK: Cambridge University Press, 2007. Available online at http://www.ipcc.ch/pdf/assessment-report/ar4/wg2/ar4-wg2-chapter14.pdf Contains information on how climate change could affect North America, including agriculture.

Gold, Mark. *The Global Benefits of Eating Less Meat: A Report for Compassion in World Farming Trust.* Petersfield, Hampshire, UK: Compassion in World Farming Trust, 2004. Available online at http://www.ciwf.org/publications/reports/The_Global_Benefits_of_Eating_Less_Meat.pdf Contains a variety of information regarding meat consumption, including conversion ratios for the amount of animal feed needed to grow each pound of meat, and the pounds of useable protein obtained per acre for various foods.

Koeppel, Dan. *Banana: The Fate of the Fruit that Changed the World.* New York: Hudson Street Press, 2007. Contains information about Panama's Disease, the fungus that threatens banana plantations worldwide.

Matthews, Roger. *The Archaeology of Mesopotamia: Theories and Approaches.* London: Routledge, Taylor & Francis Group, 2003. Contains information on the early development of agriculture, and especially on the climate forces that made it possible.

Paarlberg, Don and Philip Paarlberg. *The Agricultural Revolution of the 20th Century.* Ames, IA: Iowa State University Press, 2000. Documents the vast changes in agriculture during the 20th century.

Pimentel, David, Paul Hepperly, James Hanson, David Douds, and Rita Seidel. "Environmental, Energetic, and Economic Comparisons of Organic and Conventional Farming Systems." *BioScience* 55:7 (July, 2005): 573-82.
Contains information on how organic farming practices reduce carbon emissions, as well as how organic production outputs compare with those of conventional systems.

Pirog, Rich and Andrew Benjamin. *Checking the Food Odometer: Comparing Food Miles for Local versus Conventional Produces Sales to Iowa Institutions.* Ames, Iowa: Leopold Center for Sustainable Agriculture, 2003.
Available online http://www.leopold.iastate.edu/pubs/staff/files/food_travel072103.pdf
Contains information on the distances that local and non-local foods travel to the consumer.

Pirog, Rich, Timothy Van Pelt, Kamyar Enshayan, and Ellen Cook. *Food, Fuel, and Freeways: An Iowa Perspective on how far Food Travels, Fuel Usage, and Greenhouse Gas Emissions.* Ames, Iowa: Leopold Center for Sustainable Agriculture, 2001.
Available online at http://www.leopold.iastate.edu/pubs/staff/ppp/food_mil.pdf
Contains information on fuel use for local and non-local foods.

Reilly, J., F. Tubiello, B. McCarl, D. Abler, R. Darwin, K. Fuglie, S. Hollinger, C. Izaurralde, S. Jagtap, J. Jones, L. Mearns, D. Ojima, E. Paul, K. Paustian, S. Riha, N. Rosenberg, and C. Rosenzweig. "U.S. Agriculture and Climate Change: New Results." *Climate Change* 57 (2003): 43-69. Contains information on how Climate Change could affect agriculture in the United States.

Robinson, Jo. *Pasture Perfect: The Far-Reaching Benefits of Choosing Meat, Eggs, and Dairy Products from Grass-Fed Animals.* Vashon, WA: Vashon Island Press, 2004. Contains some good information on various reasons why grass-fed products can benefit the environment, farmers, consumers, and even the animals themselves. That includes some nutritional benefits of grass-fed products.

Shiva, Vandiva. "Globalization and the War Against Farmers and the Land." In *The Essential Agrarian Reader: The Future of Culture, Community, and the Land,* edited by Norman Wirzba, 121-139. Washington, D.C.: Shoemaker and Hoard, 2004. This chapter of *The Essential Agrarian Reader* contains some good information on the vulnerability of monocultures and industrial farming. It also contains some information on the total output of monocultures in comparison to polycultures.

Solbrig, Otto T., and Dorothy J. Solbrig. *So Shall You Reap: Farming and Crops in Human Affairs*. Washington, D.C.: Island Press, 1994.
Contains information on the development of agriculture in human history.

Steinfeld, Henning, Pierre Gerber, Tom Wassenaar, Vincent Castel, Mauricio Rosales, and Cees de Haan. *Livestock's Long Shadow: Environmental Issues and Options*. Rome: FAO, 2006.
Available online at http://www.virtualcentre.org/en/library/key_pub/longshad/A0701E00.htm
A report from the LEAD (Livestock, Environment and Development) initiative of the United Nations' Food and Agriculture Organization (FAO).
Contains detailed information on how the livestock industry is affecting the global environment.
Chapter 3 especially pertains to climate change, but other sections including the summary, introduction, and conclusion are also relevant to the issue.

Union of Concerned Scientists. "Industiral Agriculture: Features and Policy." Union of Concerned Scientists. http://www.ucsusa.org/food_and_environment/sustainable_food/industrial-agriculture-features-and-policy.html
Contains some good information on some of the dangers of industrial agriculture. It discusses how we are losing the genetic base in our crops and therefore creating a susceptibility to diseases and reducing the ability to adapt.

Willer, Helga and Minou Yussefi, ed., *World of Organic Agriculture: Statistics and Emerging Trends 2007.* Rheinbreitbach, Germany: Medienhaus Plump, 2007.
Available online at http://orgprints.org/10506/01/willer-yussefi-2007-p1-44.pdf
Documents the growing industry of organic farming around the world.

Web Resources

The website of the United States Department of Agriculture's National Organic Program. http://www.ams.usda.gov/nop/indexNet.htm

The website run by Jo Robinson focusing on grass-fed meat and dairy products. It has a lot of information on the subject, as well as a directory of farms around the country featuring this type of food.

http://www.eatwild.com/

The website of the nonprofit organization FoodRoutes. They concentrate heavily on promoting local food consumption. The site has a wide array of information and resources.

http://www.foodroutes.org/

The Local Harvest website. Contains a database of stores, farms, CSA's, farmers markets, and restaurants around the country that feature local and organic foods. Also has a newsletter, a weblog, and other information.

http://www.localharvest.org/

The website of the New Farm section of the Rodale Institute. The site is a global exchange of progressive farming knowledge.

http://www.newfarm.org/

A website of four collaborating groups; the Foundation of Ecology and Agriculture (SOEL), the International Federation of Organic Agriculture Movements (IFOAM), the Research Institute of Organic Agriculture (FiBL), and NürnbergMesse, the organizers of the BioFach Fair. It is a supplement to the 'World of Organic Agriculture' report (referenced above), with additional information and news. It also links to the report itself.
http://www.organic-world.net/

The website of the Rodale Institute, an international research, advocacy, and aid organization for regenerative farming practices and healthy soil, food, and people.
http://www.rodaleinstitute.org/

The website of the Robyn Van En Center for CSA Resources, at Wilson College in Chambersburg, PA. Contains information on Robyn Van En, CSA's, and has a database with CSA's around North America.
http://www.wilson.edu/wilson/asp/content.asp?id=804

Interviews

Maloney, Charlie. Cologne, VA, June 22, 2007.

Radio Broadcasts

Gross, Terry. "Bananas, A Stories Fruit with an Uncertain Future." *Fresh Air.* National Public Radio. Philadelphia, Pennsylvania: WHYY, February 18, 2008. Available online at http://www.npr.org/templates/story/story.php?storyId=19097412 A radio interview with Dan Koeppel, to discuss his book *Banana: The Fate of the Fruit that Changed the World* (cited above)

ACKNOWLEDGEMENTS

I would like to thank the various people who gave their time and energy to help make this book possible. Dennis Taylor at the College of William and Mary both advised and helped edit the project. J. Timmons Roberts helped me create my college major and make the opportunity possible.

Thank you to Burr Morse and everybody at Morse Farm Maple Sugarworks. Thank you to John Garn, Mimi Gatens, Mike Benziger and everybody else at the Benziger Family Winery. And thank you to Charlie Maloney and his family at Dayspring Farm, as well as his interns- Steve, Julia, and Zach.

Thank you to Jillian Herrigel for the amazing illustrations.

Finally, thank you to my family.

ABOUT THE AUTHOR

Malcolm Lewis earned a Bachelors degree in Environmental Writing from the College of William and Mary in Williamsburg, Virginia in 2008. He now lives on the coast of Maine where he works as a carpenter. In his free time he enjoys sailing, running, biking, swimming, and cooking.

In order to honestly address climate change, and the many other environmental problems we face, Malcolm promotes the idea that our species must drastically reduce both consumption and population.

"Some things are really necessaries of life in some circles,
the most helpless and diseased,
which in others are luxuries merely,
and in others still entirely unknown."

-Henry David Thoreau,
From Walden

A Note on Type

This text of this book is set in Garamond.
It is the style of text used in
The Lorax, by Dr. Seuss.